WEYERHAEUSER ENVIRONMENTAL BOOKS

William Cronon, Editor

The Rhine

An Eco-Biography, 1815–2000

MARK CIOC

University of Washington Press

SEATTLE AND LONDON

For my father, Charles J. Cioc

The Rhine by Mark Cioc has been published
with the assistance of a grant from the Weyerhaeuser
Environmental Books Endowment, established
by the Weyerhaeuser Company Foundation, members of the
Weyerhaeuser family, and Janet and Jack Creighton.

Library of Congress Cataloging-in-Publication Data

Cioc, Mark
The Rhine : an eco-biography, 1815-2000 / Mark Cioc.
p. cm. —(Weyerhaeuser environmental books)
Includes bibliographical references and index.
ISBN 0-295-98254-3 (alk. paper)
1. Human beings—Effect of environment on—Rhine River.
2. River engineering—Rhine River. 3. Rivers—Rhine River—Regulation.
4. Stream ecology—Rhine River. 5. Rhine River—History.
6. Rhine River—Environmental conditions. I. Title. II. Series.
GF540.C56   2002
333.91'62'09434—dc21   2002072694

The paper used in this publication is acid-free and
recycled from 10 percent post-consumer and at least 50 percent
pre-consumer waste. It meets the minimum requirements of
American National Standard for Information Sciences-Permanence
of Paper for Printed Library Materials, ANSI z39.48-1984.

The Rhine is the river about
which all the world speaks but no one studies,
which all the world visits but no one knows,
which one sees as it passes but forgets as it flows,
which everyone skims but no one plumbs.
Still, its ruins lift the imagination
and its destiny preoccupies serious minds;
and below the surface of its current,
this admirable river reveals to the poet and
statesman alike the past and future of Europe.

—Victor Hugo (1845)

Contents

Abbreviations

BASF   Badische Anilin- und Soda-Fabrik

CHR   International Commission for the Hydrology of the Rhine Basin (same as KHR)

IAWR   International Working Group of the Waterworks of the Rhine Basin

ICPR   International Commission for the Protection of the Rhine (same as IKSR)

IKSR   Internationale Kommission zum Schutze des Rheins gegen Verunreinigungen (same as ICPR)

KHR   Internationale Kommission für die Hydrologie des Rheingebietes (same as CHR)

LAWA   Working Group of the German Federal States

LHAK   Landeshauptarchiv Koblenz

NRW/HSA   Nordrhein-Westfälisches Hauptstaatsarchiv

OSPAR   Oslo and Paris Treaties (Commission)

PCB   polychlorinated biphenyl

RIWA   Association of Rhine and Meuse Water Supply Companies

RWE   Rheinische-Westfälische Elektrizitätswerk

Time and the River Flowing

William Cronon

Rivers are hardly a new subject for European or world history. Narratives of the earliest civilizations have more often than not centered on watercourses whose very names have become the stuff of legend: the Tigris and the Euphrates, the Yellow and the Yangtze, the Nile, the Jordan, the Ganges. European exploration and imperial expansion in the past half millennium more often than not have hinged on access to the great rivers that served as highways into the interiors of most major continents, from the Amazon to the Congo to the Mississippi. And in Europe itself, one cannot write the history of certain cities or nations without reference to the rivers that nurtured their growth. The Thames and London, the Seine and Paris, the Danube and Vienna are only the best known among scores of other examples.

And yet, when one looks more closely at the roles these rivers have played in traditional histories, it quickly becomes apparent that all too often they serve only as the stage upon which various human dramas have been acted out. Although a river may provide geographical opportunities to set in motion the growth of a great city, although it may offer the corridor along which far-reaching trade networks emerge, although it may even appear to carry the very lifeblood of an emerging nation or empire, its own history can remain surprisingly invisible. Indeed, the river is sometimes the one feature in the human landscape that seems to change very little, if at all. Useful primarily as a narrative device for linking together places and events that might otherwise seem disconnected from each other, the river—like too much else in the natural world—remains static and historyless in such accounts.

This is why Mark Cioc's *The Rhine: An Eco-Biography* is such a welcome departure in offering the first true environmental history of a major European river. One basic premise of this book will be news to no one:

the Rhine is arguably the single most important river in the history of northwestern Europe. It has played an essential role in the stories of several nations—Switzerland, the Netherlands, France, and most especially Germany—whose significance to the history of western Europe can hardly be overstated. The importance of the Rhine to European industrialization, to the contested boundaries of European nationalism, even to European warfare, has long been recognized by every scholar who studies the political and economic history of the continent. Historical geographers have done a superb job of analyzing the river both as a transportation route and as a crucial natural resource for the changing economic activities that have grown up alongside it. All of this is well known.

What Cioc now adds to this familiar material is the history of the river itself, and the result suggests the benefits to be had by adopting an environmental approach even to the most familiar historical subjects. Environmental historians have long argued for a dialectical understanding of the changing interactions between human beings and the natural world. Against those who write as if history can be understood purely in human terms, they point to the nonhuman creatures and systems that provide the material contexts for human lives and often play crucial roles in changing the course of human events. Against those who write of nature as if humans have no place in it, or who view history primarily in biological terms, they point to the importance of human ideas and human cultural conventions not just in reshaping natural systems, but in providing the very language and concepts with which we understand those systems. Like other members of history's guild, environmental historians are committed to the basic premise that things change, and the dynamism of nature is as intriguing to them as the dynamism of human society.

Building on these intellectual foundations, Mark Cioc offers us a Rhine river that is far more dynamic than past histories might lead us to believe. On the one hand, it became the means whereby Germany rose to industrial greatness in the nineteenth and twentieth centuries. It supplied the water without which the giant (and highly polluting) chemical factories lining its banks would have been impossible, and it carried the ships whereby those same factories obtained their supplies and sent their products to market, carrying away the factories' effluent at the same time. Ironically, its upper reaches also served, at least in the early years of industrialization, as one of the chief icons of European romanticism, a powerful aesthetic symbol of nonhuman nature festooned with architectural monuments of a Gothic past that became a destination for travelers seeking escape from what they saw as the gritty squalor of modernity. Different visions of what makes the Rhine "useful" and "beautiful" have competed with each other for at least the past two centuries, and these visions are among the most important engines of change in Cioc's story.

The Rhine in this book is far from being a passive stage upon which the march of human progress takes place. Yes, the river has been profoundly altered by the efforts of engineers and planners to "rationalize" its natural vagaries. And the reinvention of the river's nature is the most intriguing and troubling story Cioc has to tell. The Rhine's emergence as Europe's preeminent commercial river was hardly the work of nature alone. Generations of planners, industrialists, and civil engineers all contributed to this achievement. Together, they straightened the river's channel, constrained its floodplain, regulated its flow, and profoundly manipulated its ecosystems, all with the goal of making it obey human visions of safety, efficiency, and productivity. Assuming that meanders and multiple channels and floodplains were all examples of natural "waste," they eliminated such waste and in the process sculpted what sometimes seemed an entirely new river.

At the time, the value of these transformations seemed self-evident to many. Who could possibly object to a safer and more productive river? But as so often happens, history's law of unintended consequences eventually demonstrated that earlier visions of "improvement" had been based on inadequate understandings of ecology and hydrology alike. A few prescient scientists like Robert Lauterborn began to document what was being lost even before the end of the nineteenth century. Vanished oaks, salmon, and wetlands were only the most visible of the environmental degradations created by too narrow a definition of "progress." Rising quantities of heavy metals and other toxic substances in the river's water imperiled human and nonhuman life alike, while also creating problems for the very industries that had proliferated those poisons in the first place. Shifts in temperature altered ecological communities in still subtler ways. Perhaps most strikingly of all, efforts at flood control and channelization accelerated the river's flow, changed its sediment loads, and began to threaten the survival of its delta, so that human actions as far upstream as Switzerland began to cause problems for human communities as far downstream as the Netherlands. Scarcely anyone had foreseen such consequences, but all were evidence that the river's nature was not to be so easily subordinated to human will as the engineers had once imagined.

The story Mark Cioc tells in these pages is both fascinating and cautionary. What happened to the Rhine during the nineteenth and twentieth centuries was unquestionably one of the success stories of modern history, a triumphant example of the benefits to humanity that can flow from harnessing nature's power. But an immense price was paid for those benefits, so much so that by the closing decades of the twentieth century, enormous efforts were being expended to recover some of the creatures, ecosystems, and natural processes that had once been sacrificed to the dream of progress. A more chastened vision of improvement now looked to nature itself as a source of authority to constrain human ingenuity. What Francis Bacon had declared at the dawn of the scientific and technological revolution had been proven true even by a river that had seemingly been sacrificed to that revolution. "Nature, to be commanded," Bacon wrote, "must first be obeyed." If we are to achieve the kind of understanding both of nature and of ourselves that such an obedience requires, surely histories like this one will be among our most profound and challenging guides.

Acknowledgments

The idea of writing this book first occurred to me while I was cycling in Europe several years ago. Many of Europe's bike routes are converted towpaths—relics of an era when animal muscle was needed to haul boats and barges upstream—and their close proximity to river channels makes them ideal for observing what has happened to riparian corridors over the past two centuries. I embarked on a short and relaxing trip but found myself on a long and pleasant research journey instead.

I have much more than my tire-repair kit to thank for the completion of this book. I am especially grateful to Rainer Doetsch, director of the Rhein-Museum in Koblenz, for allowing me unrestricted access to the museum's superb collection of books, documents, and artifacts. Without his generosity, I would not have been able to locate my sources with such ease and enjoyment. I am also deeply indebted to Bruce Thompson, my colleague at the University of California, Santa Cruz, for his unflagging willingness to read many different versions of this manuscript over the past three years. Every page has benefited from his intellectual acumen and encyclopedic knowledge. I am equally indebted to J. R. McNeill and William Cronon for their gentle but penetrating critiques of the all-too-rough draft I submitted to the University of Washington Press back in 2000. They helped me to think through the implications of my research and to frame my story within a larger perspective. Special thanks to Julidta Tarver for guiding the manuscript through the publication process. Thanks also to Betsy Wallace and Marnia He-Sapa for making the text more accessible to readers outside the field of history. Finally, I wish to thank my father, cartographer *extraordinaire*, for his assistance in drawing the maps and graphs.

Books are never authored alone, and I would like to thank everyone who has helped me along the way.

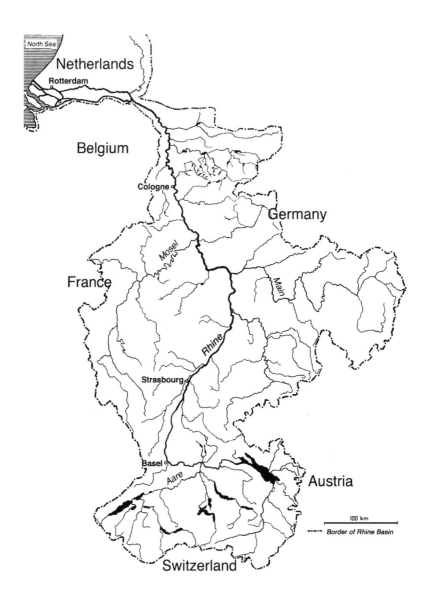

Figure 1.1 The Rhine river basin (Adapted from Wasser- und
Schiffahrtsdirektion Duisburg, *Der Rhein: Ausbau*, 35)

The Rhine

I love rivers. They carry ideas
as well as merchandise.... Rivers, like clarions,
sing to the ocean about the beauty of earth,
the cultivation of fields, the splendor of cities,
and the glory of humans.
And of all rivers, I love the Rhine most.

—Victor Hugo (1845)

My Rhine is dark and brooding.
It is too much a river of merchant cunning for me
to believe in its youthful summertime face.

—Heinrich Böll (1960)

CHAPTER 1

Introduction

*T*he modern Rhine—"Europe's romantic sewer"—is an offspring of the French and industrial revolutions. Conceived by Napoleon and designed by engineers, the river acquired its canal-like profile during the nineteenth century. Three events in rapid succession marked its birth. In 1815 the Congress of Vienna placed the Rhine under an "international regime" designed to accelerate the free flow of trade. In 1816 the first Rhine steamer, *Prince of Orange*, chugged upstream from Rotterdam to Cologne, inaugurating the age of coal and iron. Then, in 1817, the Baden engineer Johann Gottfried Tulla began the most ambitious rectification work ever undertaken on a European river. Celebrated as the "Tamer of the Wild Rhine," Tulla is best known for the simple maxim that guided his work: "No stream or river, the Rhine included, needs more than one bed; as a rule, multiple branches are redundant."[1] Cooperation, coal, and concrete: together they started a riparian revolution that has determined Rhine affairs ever since.

None of the Vienna delegates had any inkling as to the real significance of what they had just created. All they meant to do was foster trade among the riparian states after twenty-five years of war and bloodshed. To this end, they established the Central Commission for Rhine Navigation (the Rhine Commission) and gave it the task of eliminating the river's commercial chokepoints—human ones, such as the innumerable toll booths, and natural ones, such as the Bingen reef and Lorelei cliffs—which had hindered river traffic for centuries. "The Rhine can count more tolls than miles," went a popular rhyme of the time, "and knight and priestling block its path."[2]

Placing the river in the foster care of the Rhine Commission proved a mixed blessing. On the positive side, the new river regime stimulated economic growth and free trade on its banks. The Rhine is today one of the world's greatest commercial arteries, in volume of traffic second only to the Mississippi. It transports millions of tons of coal, steel, chemicals, pharmaceuticals, textiles, and other goods each year, many of which are produced directly on its banks. The Rhine Commission, now headquartered in Strasbourg, can justifiably lay claim to being the oldest continuous interstate institution in Europe and the first step in the long march of diplomacy that culminated in the Common Market and European Union.

On the negative side, the multinational engineers who took possession of the river in 1815 were strict disciplinarians, whose idea of a well-behaved river was not a river at all: it was a canal, utterly and completely harnessed to the needs of transport. They did not view themselves as custodians of the Rhine's fish stocks and alluvial forests, although salmon and timber were the mainstays of river commerce at the time. Nor did they see themselves as protectors of the Rhine's broad floodplain, although it was an integral part of the river's drainage system and home to a rich variety of flora and fauna. The birth of the new Rhine thus spelled doom for the old one. First, engineers severed the river's arms and braids from its trunk as dictated by the Tulla maxim. Then industries and cities introduced slow-acting poisons into its water system. The result was a truncated river shorn of its biological diversity.

This book traces the life story (or "biography") of the Rhine from 1815 to 2000. It focuses on how and why the river became a degraded biological habitat, and on the attempts since the 1970s to resuscitate and nurse it back to health. The entire river—from its high headwaters in the Swiss Alps to its muddy delta in the Netherlands—forms the subject of this study. The main channel, tributaries, floodplain, islands, and underground flow are all treated as parts of the Rhine, as are the life forms it sustains. Humans are the principal actors. Sometimes they appear as representatives of one of the riparian states: Switzerland, Austria, Liechtenstein, Germany, France, Belgium, Luxembourg, the Netherlands. More often they are seen in their role as engineers, entrepreneurs, water experts, biologists, fishermen, politicians, or diplomats, since each group tended to think along similar lines despite differences of time and place. Industrialists everywhere on the river spoke the language of economic progress and environmental laissez-faire. Germany's urban planners faced the same clean-water dilemmas as their counterparts in France. Swiss and Dutch fishermen alike lost their livelihoods.

National rivalries and warfare enter the story only when they make a palpable difference to the fate of the river, most obviously in Alsace and Lorraine, the two contested provinces situated on the Rhine's left bank. Alsace and Lorraine, however, were the exception. The international river idea has survived remarkably well for nearly two hundred years, even if it was (until 1945) mauled occasionally between the jagged teeth of France and Germany. That General George Patton publicly urinated in the Rhine to display his contempt for Nazi Germany meant nothing to the river's health. That millions of Europeans, before and after Patton, regularly flushed their raw sewage down the river meant everything. Round-the-clock Allied aerial bombing during World War II had a relatively short-term impact on the river's life, for the Rhine bridges and harbors were quickly repaired at war's end. But continual assaults from Rhine-based coal, steel, and chemical industries have left the river's ecology crippled and disfigured.

Anyone familiar with the Danube, Mississippi, Hudson, Donets, and other major "industrial rivers" will instantly recognize the general outlines of this story: the Rhine Commissioners set out to manipulate and control the river as fully as possible (to "tame," "train," "rectify," "ameliorate," "straighten," and "improve" it in their terminology), only to find themselves caught in a long war of attrition. When engineers closed off the Rift valley floodplain in an effort to protect the upstream cities of Strasbourg and Ludwigshafen, the river began to inundate the downstream cities of Koblenz and Nijmegen instead. When humans in their folly depleted the Rhine's savory salmon, shad, and sturgeon stocks, the river served up the less palatable roach, bleak, and bream in their place. When industries overwhelmed the riverbed with heavy metals, the Rhine spat them back undigested into drinking water supplies and onto irrigated agricultural fields. Not until the 1970s did the riparian states begin to comprehend the extent of the damage they had wrought in a war they did not want to win. The vast sums of money now being pumped into salmon repopulation, floodplain restoration, toxic-waste cleanup, and water purification are really nothing but reparation payments for two centuries of inadvertent ecocide.

The notion that a river is a biological entity—that it has a "life" and a "personality" and therefore a "biography"—is not altogether out of step with scientific or commonsense notions of rivers. Rivers seem alive to us—restless, temperamental, fickle, sometimes raging, sometimes calm. They are forever on the move, collecting atmospheric precipitation and transporting it back to the earth's basins as part of the global water cycle. Gravity and sunlight lend energy to rivers, making them active sculptors of the landscape. They chisel away at mountains, grind boulders to sand, and etch floodplains through the Earth's crust as they transport rock and sediment downstream. On occasion, they also overspill their banks and pour into shops and cellars, wreaking havoc like vandals in the night.

Many descriptions betray an underlying sense that rivers etch their own signatures on the landscape, that they carve out unique profiles. Scientists use terms such as "young," "mature," and "senescent" to depict a river's life cycle. "Dead river" refers not to a stream that has dried up but to a flowing stream that supports no life. Rivers have a kind of "metabolism." A century ago, most Europeans still believed that the "self-cleansing" capacity of rivers—their astonishing ability to absorb and neutralize vast quantities of toxins and wastes—derived from the mechanical cleansing action of underwater rocks and cliffs. Now it is well known that rivers remain clean primarily through the biological activity of bacteria. Rivers, of course, are not actually alive. That is but a literary conceit. But they do carry the single most important substance for the maintenance of life on earth: *water*. They lie at the conflux of the physiochemical and biological worlds, providing a living space for fish, snails, insects, birds, trees, and people. Richard White's memorable tag for the Columbia is applicable more generally: every river is an "organic machine."[3]

Humans live on rivers for much the same reason that other organisms do. Rivers provide a ready supply of nourishment and a convenient mode of transport. And any river with human inhabitants is much more than just a physical and biological entity: it is also the site of political, economic, and cultural activity. The Rhine's historical identity is inexorably intertwined with thousands of years of human culture, human labor, and human manipulation. Its earliest known inhabitants include Heidelberg man, named after the Rhine city closest to the first excavation site, and the Neanderthals, named after the Rhine's Neander Valley (near Düsseldorf) where the first skull was found. It was the ancient Celts who bequeathed the river its name, *Renos*, from which all subsequent variations are derived: *Rhenus* (Latin), *Rein* (Rhaeto-Romansch), *Rhein* (German), *Rhin* (French), *Rijn* (Dutch), *Rhine* (English). By the time the Romans arrived, more than two thousand years ago, the Rhine was already a densely populated, multicultural basin: "The Rhine rises in the land of the Lepontii, who inhabit the

Alps," wrote Julius Caesar in the oldest known historical text discussing the Rhine, *Commentarii de bello Gallico*. "In a long swift course it runs through the territories of the Nantuates, Helvetii, Sequani, Mediomatrices, Triboci, and Treveri, and on its approach to the Ocean divides into several streams, forming many large islands (a great number of which are inhabited by fierce barbaric tribes, believed in some instances to live on fish and birds' eggs); then by many mouths it flows into the Ocean."[4]

Of all the cultures of antiquity, Rome's legacy is still most visible today, mainly in the form of city names and vine-clad towns, such as Chur (*Curia Rhaetorum*), Basel (*Basilia*), Bacharach (*Bacchi Ara*), Mainz (*Mongontiacum*), Cologne (*Colonia*), and Nijmegen (*Noviomagus*). But the "Roman Rhine" is only one of the river's many cultural overlays. By the Middle Ages, the Rhine was better known as "Priest Street," a derisive tribute to the importance of the Mainz, Cologne, and Trier archbishoprics in the election of Holy Roman Emperors. When the Dutch revolted against the Habsburgs in the sixteenth century, the Rhine became infamous as the "Spanish Road," the military supply route that linked Spain to its possessions in the southern Netherlands (Belgium). And castle ruins, strewn everywhere on the hillsides between Strasbourg and Cologne, serve as testimonies in stone and mortar to the battles between French and German-speaking peoples—the Thirty Years' War, the campaigns of Louis XIV, the French Revolutionary and Napoleonic wars, World Wars I and II—that have punctuated European history over so many centuries.

The "Romantic Rhine"—the stretch between Mainz and Cologne full of quaint villages, terraced vineyards, and winding gorges—is one of the most enduring of the river's many images. Ostensibly rooted in the Rhine's hoary past, in truth it was the invention of Dutch and British "ecotourists" of the seventeenth and eighteenth centuries. Traveling to and from Italy, notebook and paintbrush in hand, they were seduced by the river's rich symbolic geography—the Mouse Tower, Sankt Goar, Drachenfels, the Lorelei—and collectively they etched the words "sublime" and "picturesque" in indelible mist and fog across the Rhenish Slate Mountains. "The Rhine nowhere, perhaps, presents grander objects either of nature, or of art, than in the northern perspective from Sankt Goar," wrote the Gothic novelist Ann Radcliffe after hiking up to the Rheinfels castle. "There, expanding with a bold sweep, the river exhibits, at one *coup d'oeil*, on its mountain shores, six fortresses or towns, many of them placed in the most wild and tremendous situations; their ancient and gloomy structures giving ideas of the sullen tyranny of former times."[5] So firmly did Radcliffe and others fix the word "Romantic" in the mind's eye that subsequent generations felt free to wax poetic about the river, whether they had bothered to visit it or not. "It was like a dream of the Middle Ages," wrote one famous non-visitor, Henry David Thoreau, after glimpsing a Rhine painting in 1862. "I floated down its historic stream in something more than imagination, under bridges built by the Romans, and repaired by later heroes, past cities and castles whose very names were music to my ears, and each of which was the subject of a legend."[6]

Just as humans have stamped their own cultural templates onto the Rhine, so too have they projected human features onto it, more so than on any other river in the world except perhaps the Nile. "Your Highness Rhine, my sweet dreams / How can I sing your praise?" began Joost van den Vondel's poem "De Rynstroom" (1629), progenitor of a whole genre of modern Dutch river elegies.[7] "A poet's dream," declared Heinrich von Kleist about the Rhine, "which now opens, now closes, now blooms, now is desolate, now laughs, now alarms."[8] For Friedrich Hölderlin, the "free-born" Rhine was the fluvial incarnation of Rousseau, and he asked rhetorically:

> But where is the man
> Who can remain free
> His whole life long, alone
> Doing his heart's desire,
> Like the Rhine, so fortunate
> To have been born from
> Propitious heights and sacred womb?[9]

Lord Byron imagined he saw the "castled crag of Drachenfels" as it "Frowns o'er the wide and winding Rhine."[10] Heinrich Heine conjured up a nymph on the cliffs of the Lorelei, who lured boatmen to their death with her siren's song.[11] For Victor Hugo, the greatest of the Rhine Romantics, the river possessed all riparian virtues rolled into one: "It is rapid like the Rhône, broad like the Loire, encased like the Meuse, serpentine like the Seine, limpid and green like the Somme, historical like the Tiber, royal like the Danube, mysterious like the Nile, spangled with gold like an American river, and abounding in fables like an Asian one."[12]

Many rivers serve as political borders—boundaries that mark and reinforce cultural differences. The Rhine, however, has never served well as
a frontier, despite a centuries-long attempt by the French to make it one.
(Even today only 350 kilometers, less than one-third of its length, form a
national border.)[13] It is true that the ancient Romans generally settled on the
left bank, Germanic tribes on the right bank—a fact that Louis XIV used
as a legal fig leaf when he seized the Alsatian capital of Strasbourg in 1681 as
part of his campaign to extend France's eastern frontier "jusqu'au Rhin."[14]
But the boundaries of old—as Albert Demangeon and Lucien Febvre so
brilliantly showed in 1935—were far too fluid for later generations of propagandists to apportion according to their own whims.[15] There were plenty
of Roman settlements on the Rhine's right bank, especially south of the
"limes" (the fortifications linking the Rhine and Danube), which Roman
legions held for an extended period. By the fourth century A.D., moreover,
the roles had reversed and it was the Germanic tribes that were crossing the
river in successive waves, pushing out the Romans and taking over—permanently—their left-bank cities and territories.

Yet if the Rhine has never functioned well as a political or cultural
border, it also never came under the exclusive suzerainty of any linguistic
grouping—Latins and Germans included—at least not long enough to
have left an indelible *ethnic* stamp on the river. It has, in other words, always
been an international stream. Anti-French sentiment in the nineteenth century, however, did transform the river into a potent symbol of German cultural and political unity, adding yet another layer to the Rhine's complex
identity. "Germany's river, but not Germany's border," trumpeted Ernst
Moritz Arndt as Germans rallied to defeat Napoleon in 1813.[16] "Dam the
Rhine with [French] corpses, cram it full of their broken bones," echoed
Kleist in even more militant tones.[17] When war loomed anew in the 1840s,
German poets and songwriters helped bring the French to the negotiating
table with a volley of verse, mustering some four hundred patriotic Rhine
tunes in the span of a single decade, a production rate that has never been
surpassed. "They shall not have it / The free German Rhine" went the
refrain of Nikolaus Becker's famous *Der freie Rhein*. "Dear Fatherland, have
no fear / The watch on the Rhine stands fast and true," proclaimed Max
Schneckenburger's equally famous *Die Wacht am Rhein* (a poem so inflam-
matory that after the Franco-Prussian war of 1870–71 Bismarck claimed
it had done the work of three divisions).[18] In 1877, Kaiser Wilhelm I even

commissioned the emplacement of a mammoth statue, *Germania*, high above the Rhine near Rüdesheim. Forged in part by melting down captured French cannons, it stands today as a grotesque reminder of Germany's nationalist pretensions on the Rhine.

This book begins in 1815, at the moment when the Congress of Vienna gave the Rhine yet another identity: that of a Euro-river. In fact, even as Arndt was proclaiming the Rhine to be "German," European statesmen were embarking on one of the boldest experiments of the nineteenth century: the internationalization of the Rhine as a commercial waterway and the establishment of a free-trade zone along its banks. Construction of the Euro-Rhine entailed a level of intervention that went far beyond anything the river had previously experienced at the hands of humans. It is true that much blood, sweat, and tears went into the construction of the "Roman Rhine" and the "Romantic Rhine," just as untold toil went into the maintenance of the river's many towpaths and terraced hillsides. Labor was a part of every person's daily encounter with the river, whether one fished in its waters, sailed against its current, or washed clothes on its shores. But at the end of the day, all that endured were castles, houses, harbors, bridges, pathways, paintings, and poems: the impact on the river itself was almost nil, its physical appearance and hydrology remaining much as before. The Euro-Rhine was different. Creating it required the use of dredgers, dynamite, dams, locks, mathematical formulas, and other tricks of the engineering trade.

Even a cursory glance at today's Rhine reveals a profile that is fundamentally different from what it was two centuries ago. The river's most celebrated features—its variegated landscape, quirky flow, and treacherous cliffs—were once the stuff of myth and legend. Now the river more resembles a canal—a monotony of barges and ore carriers—than a fabled or mysterious stream; and it flows more like an industrial faucet than a natural river. The old meandering and braided bed has been straitjacketed and streamlined into one main channel. It flows faster and with less variation in its water level than it did in the past. Its banks are now mostly lined with cement, gravel, and wing dams (groynes) instead of oak, elm, and willows. Its width, which once fluctuated wildly, is now wholly predetermined at any given point: precisely 200 meters at Basel, 300 meters at Koblenz, 400 meters at Düsseldorf, 1000 meters at Hoek van Holland. Its shipping lanes are kept at prescribed minimum depths ranging from 1.7 to 2.5 meters (incredibly shallow for one of the world's premier waterways). It has been shortened by a total of 105 kilometers, representing a loss of nearly eight percent of its original length. Gone are the thousands of islands, the braided beds, the snaking curves, the oxbows, the rivulets, the meadowlands, the fishing villages, and other eccentricities that once delighted travelers. Gone too, at long last, are the Schneckenburgers of Germany standing *Wacht am Rhein* against the perfidious French. Today the foot soldiers have been replaced by water engineers and navigation officials—German, French, Swiss, and Dutch—jointly "standing watch" over thousands of hydro-dams, locks, wing dams, harbors, depth meters, and barges. The Rhine, once free, is everywhere enchained.

"Rhenus gelidis undis" ("icy-waved Rhine") was the Roman writer Lucan's description two thousand years ago.[19] Even a hundred years ago the Rhine still froze over occasionally in winter, halting shipping for weeks at a time and turning the river into a skating rink. Now its non-Alpine stretches stay virtually ice-free year round, in part because of the salty residue from French potash mines, in part because of the heated wastewater from German power plants, and in part because so many Dutch barges ply up and down it day and night that ice has no chance to form and collect. The water table has sunk drastically in the Rift valley between Basel and Mannheim, as well as in the coal-mining regions of North Rhine–West-phalia. Silt that was once destined for the upstream floodplain now gets dredged out of the Dutch delta or poured over agricultural fields through flooding. Much of it is so laden with pollutants—cadmium, mercury, lead, nickel, antimony, chromium, zinc, salts, phenol (carbolic acid), and pes-ticides—that it cannot be used for land reclamation or farming.[20] What would Victor Hugo say about the river were he alive today? That it is pol-luted like the Danube, overpopulated like the Yangtze, harnessed like the Mississippi, overutilized like the Colorado, dammed like the Columbia, cluttered like the Hudson, straight like the Panama canal?

It is not simply that commercial fishing and crabbing have all but dis-appeared, or that spectacular catastrophes (such as the infamous Sandoz chemical spill of 1986) periodically wreak havoc on the river's flora and fauna. The Rhine of today is a fundamentally different riparian habitat from the river of 1815. The old riverbank vegetation is all but gone, as are most of the old-growth forests and all of the salmon runs. They live today only in paintings and maps on museum walls and in the collective memory of poetry and song. Of the forty-seven fish species that swam in the Rhine two hundred years ago, only about half of them could still be found there by the 1970s, and many of those survived only because of fish hatcheries. Salmon, shad, and sturgeon—the three most important commercial species—had all vanished, along with the fishermen who once earned their livelihood from them. Meanwhile, several new fish migrated to the Rhine, or were transported there—all of them less dependent on pure water, floodplains, natural riverbanks, and meadowlands than their predecessors. The same is true of invertebrate macrofauna, key indicators of water quality: in 1915, there were eighty known indigenous species, by 1956 only forty-two, by 1971 only twenty-seven. Several native crab, mussel, and snail species went extinct, their ecological niches taken by species more resistant to salts and pollution. Dozens of bird species disappeared as well.[21] "More faunal changes have occurred in the past one hundred fifty years," noted the biologist Ragnar Kinzelbach in 1984, "than in the previous ten thousand years."[22]

One hundred fifty years is a short span in a river's life, a drop in the bucket of time, a *petite durée*. The "primeval Rhine" came into being millions of years ago, a consequence of the uplift that gave rise to the mountains of Central Europe (see figure 1.2). As time passed, the Rift valley opened at the foot of the Alps (between today's Black Forest and Vosges) to become part of the river's nascent floodplain. The river, meanwhile, etched and re-etched its path northward and westward to today's North Sea, outpacing the uplift of the Rhenish Slate Mountains and giving rise to the stretch of the Rhine now most renowned for the beauty of its natural contours. The Rhine was not originally an Alpine river: the Aare and Alpenrhein flowed eastward via the Danube toward today's Black Sea. The Aare first found its way to the Rhine about two million years ago, the Alpenrhein perhaps a half million years ago. The Dutch delta, meanwhile, took on its current shape about ten thousand years ago, as a consequence of the Wurm ice age.[23]

Nineteenth-century river engineering thus marked a turning point in the Rhine's history, its greatest since the past ice age. The last time an event of this magnitude occurred on the Rhine, it still shared a common delta with the Thames.

Watershed ecosystems are an attractive subject for environmental historians because they offer spatial boundaries for investigation. Riparian communities, however, are dynamic and open, not closed and stable. Anyone

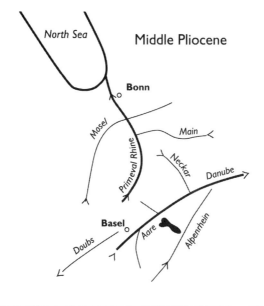

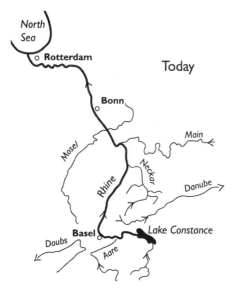

Figure 1.2 The primeval Rhine and today's Rhine (Adapted from
Tittizer and Krebs, eds., *Ökosystemforschung*, 9)

studying a river must take into account that flora and fauna migrate with or without human assistance, and that watersheds are as fluid as the substance that defines them. Heraclitus made much the same point, metaphorically, back in the sixth century B.C., and with greater brevity than do most ecologists today: "Panta rhei" ("Everything flows"). What makes the Rhine watershed a useful spatial boundary, however, has less to do with the natural instabilities of riparian habitats than with the dynamics of European political and economic life. The Rhine would look different today from the Rhine of yesteryear even if it had never become an "international river," or no steamers had entered its waters, or no engineers had tinkered with its hydrology, or indeed if all its human inhabitants had collectively moved to the Volga two hundred years ago. But it would be hard to imagine that under these circumstances the Rhine would now be virtually denuded of its islands and forests, or that salmon would have vanished from its waters. Today's Rhine, in other words, is not just a geological entity—a construct of plate tectonics, volcanic activity, climatic variation, soil erosion, and other natural processes. It is above all a human artifact, a techno-river, a thoroughly anthropomorphized stream.

Chapter 2 highlights the Rhine's international character: as a geopolitical entity flowing across linguistic and national borders; as an economic axis dividing and uniting western Europe; and as a jointly administered Euro-river promoting transnational trade and commerce. The *ancien régime* Rhine—the river as it existed politically in 1789 on the eve of the French Revolution—was a landscape of petty quarrels. Between Alsace in the south and the Netherlands in the north there were ninety-seven distinct German states alone. The "knights and priestlings" who ruled these lilliputian lands thought parochially and acted locally: they warred with their neighbors over fishing holes and birded islands, built dams that increased the number of sandbars and forks, defended their "staple" and "transfer" privileges, and manned the toll booths (thirty-four in a 600-kilometer stretch from Germersheim to Rotterdam alone)—all to the detriment of river trade.[24] By contrast, the half dozen or so Rhine states of the post-Napoleonic period thought globally and acted like cosmopolitans. They scooped out the islands and filled in the fishing holes to create a uniform navigational bed. They dismantled the toll booths and turned the river into an autobahn of world trade. They even fought their wars on a world stage. The Rhine thus passed from local hands to global hands without any intermediate institutional link between them. By placing the Rhine under the institutional control of the Rhine Commission, European diplomats unwittingly created a variant of what Garrett Hardin has called "the tragedy of the commons":[25] all of the riparian states had a vested interest in maximizing their share of the Rhine's commerce and trade, while none felt any real responsibility for preserving it as a riparian habitat. All would come to bemoan the degradation of the river, yet none had any strong incentive to do anything about it unilaterally.

Chapter 3 provides an in-depth look at the great nineteenth- and twentieth-century engineering projects that gave rise to the Rhine's modern profile. Despite the Rhine Commission's efforts to control the pace and scope of the rectification work, there was an unmistakable "sorcerer's apprentice" quality to the way the river was manipulated and transformed. The engineering blueprints spelled out in great detail the intended goals and objectives: flood control, irrigation, drainage, ease of navigation, hydroelectric power, a steady water supply, and so forth. Yet rarely did the engineers correctly anticipate the full consequences of their manipulative actions, and more often than not they were caught off guard when the river responded in unexpected ways. Invariably this meant another flurry of blueprints and projects designed to correct the previous ones. Adding to the engineering difficulties were changes in the political goals over time. Tulla focused on flood control measures at the base of the Alps, where most of the river's worst floods occurred. Subsequent projects by later generations of engineers highlighted navigation, not flood control, especially once steam power began to be exploited to its fullest extent after 1850. Changes in objectives over time meant that some stretches of the river were revisited and reconstructed many times. Techno-fixes begat techno-fixes until the river was straitjacketed more than any large river before or since.

The Rhine's modern contours were designed and crafted by engineers, but the projects were largely tailored to the needs of the nineteenth-century European economy, especially Germany's coal and chemical industries, the subject matter of chapters 4 and 5. The advent of "carboniferous capitalism,"[26] to borrow Lewis Mumford's phrase for the age of fossil fuels, was prefigured millions of years ago, when organic material was deposited and compressed into coal along a broad band of northwestern Europe that included the Rhine and Ruhr. Coal from Prussia's two Rhine provinces—Westphalia and Rhineland—fueled the powerful military-industrial juggernaut that Bismarck used to forge German unification in 1871. Coal-ore exports and iron-ore imports dictated the kind of engineering done on the lower stretches of the river and its tributaries, especially the Ruhr, Emscher, Lippe, and Erft. Coal dust, phenol, and other mining-related pollutants created serious water pollution problems on the Rhine and Ruhr, while acid rain from coal smoke ate away the surrounding forests and vineyards, creating human health problems of breathtaking proportions. And coal-tar derivatives, used in the manufacturing of the first synthetic dyes, gave rise to the behemoth German and Swiss chemical plants on the river—still today the driving force behind the Rhine's economy and one of the main sources of river pollution.

Chapter 6 examines the long-term consequences of Rhine engineering and economic development on the Rhine's biotic communities. In 1817, the year the Tulla Project began, the Rhine was enveloped in nearly 2,300 square kilometers of floodplain (more if the Alpine feeder streams are added to the tally).[27] This floodplain formed a continuous band stretching along a thousand-kilometer corridor from Lake Constance to the Dutch delta. It varied in width depending on the terrain—in some places only a few hundred meters, in others as wide as fifteen kilometers. This corridor of forests, meadows, marshes, and reeds gave the Rhine its geographic breadth, its biological diversity, its ecological dynamism, and much of its self-cleansing capacity. By 1975, however, nearly nine-tenths of this floodplain had been usurped by farmers, industrialists, and urban planners. The remaining one-tenth, too meager to function as a continuous corridor, consisted of small patches in a riverscape otherwise dominated by human enterprise. With the alluvial corridor went the living space upon which the river's nonhuman inhabitants depended for their survival, resulting in the loss or near-loss of beaver, otter, salmon, shad, sturgeon, plover, sandpiper, tern, dunlin, heron, tree frog, dice snake, Rhine mussel, mayfly, oak, ash, elm, willow, and hundreds of other animal and plant populations.

Chapter 7 examines current efforts to restore the river's health. Much of the rehabilitation work is being carried out by a new organization, the International Commission for the Protection of the Rhine against Pollution (the Rhine Protection Commission), which since 1963 has come to overshadow the original Rhine Commission. What drove the riparian states to reassess their Rhine regime was, above all, the recognition that the river could no longer handle the multitude of roles assigned it in the vast industrial-agricultural-urban-tourist nexus that had grown up on its banks. The river could not function as a conduit for industrial and agricultural wastes and still provide clean water to cities; it could not support endless urban sprawl and still be a favored destination for tourists; and it could not offer safety to anyone as long as it repeatedly overspilled its artificial banks.

As the new organization's name suggests, pollution abatement rather than habitat restoration was uppermost in the minds of those who created it, at least initially. But the 1986 Sandoz chemical spill did much to awaken a greater interest in preserving the river as a biological habitat. The progress to date—which can be seen in the form of cleaner water, biotic recovery, and floodplain restoration—has brought enormous benefits to humans and nonhumans alike. The most celebrated of the projects now under way, "Salmon 2000," has even managed to coax a few salmon back to the river after an absence of nearly fifty years. There are, however, limits on what the new commission can accomplish now that millions of Europeans are wholly dependent on the river's goods, water, and energy. Full "restoration" is an impractical goal. All that the new commission can hope to achieve is a partial resuscitation—enough to allow the river of old to show its face here and there amid the harbors, hydrodams, factories, and cities.

Too often economic historians see only the steamship's forward progress, while environmental historians see but the smokestack's plume. Depending on perspective, a fishing village can be depicted as pleasingly arcadian or hopelessly backward, a factory as benignly progressive or rapaciously exploitative. Such simple abstractions, however, mask more than they reveal: rarely are the lines separating progress and decline, good and evil, winners and losers, so easily drawn. The fisheries were the most obvious losers on the new Rhine. No amount of pisciculture could save them in the face of round-the-clock dam construction, gravel extraction, bed straightening, and water pollution. Yet fishermen hastened their own demise by overfishing the shad and salmon: no fish treaty stopped them from depleting and eventually annihilating the Rhine stocks. Moreover, old fishing villages such as Sankt Goar and Bacharach did not simply dry up and wither away as the fish disappeared; they turned to the highly profitable business of angling tourists with the "Romantic" bait.

Similar if less painful dilemmas faced everyone who lived, worked, and traveled on the Rhine as it was redesigned to fit the needs of industrial Europe. Rhine farmers gained from land reclamation but lost from a sinkage in the water table, both of which were consequences of river rectification. They gained when the chemical industry developed new synthetic fertilizers but lost when chemical pollutants seeped into their irrigation channels. Some laborers traded the backbreaking task of tugging barges for the slightly less onerous one of stoking steam engines. Others found jobs, for better or worse, in the burgeoning iron, steel, sugar-beet, pulp-and-paper, and chemical plants. Still others spent their lives doing the rectification work itself—dredging the channel, removing the gravel, reinforcing the banks, constructing harbors, and the like. Visitors continued to romance the river throughout the nineteenth century, even as they saw first-hand the transformations taking place. In fact, they came in droves. Their preferred modes of transportation were Rhine steamers and locomotives; their preferred places to sleep were riverbank hotels with all the modern conveniences; their preferred food, salmon and shad.

There were, of course, Cassandras. The English Romantic, Samuel Taylor Coleridge, in a famous verse, saw nothing but trouble ahead:

> The river Rhine, it is well known,
> Doth wash your city of Cologne;
> But tell me, Nymphs, what power divine
> Shall henceforth wash the river Rhine?[28]

Fritz André and a few other river engineers warned as early as the 1820s that upstream rectification work would cause an increase in downstream flooding. In 1857, George Meredith composed a Rhine horror tale set in Cologne that pitted the hero "Ferina" (a reference to the town's famous perfume, Eau de Cologne) against the evil "yellow devil" (a reference to the sulfur dioxide belching out of the town's new industrial furnaces). Swiss and German nature groups tried unsuccessfully to thwart the construction of a hydro-dam at Laufenburg, site of the Rhine's most spectacular salmon runs, in the early twentieth century. But these voices were largely lost amid the dint and clatter of the steam age, not least because cause and effect were often difficult to establish: typically the economic benefits came immediately, whereas the ecological damages took generations to appear.

Ultimately, the new Rhine was constructed not just by Eurocrats and engineers but also by the millions of people who lived and traveled on its banks every day—by laborers who preferred the tugboat to the towline, by boaters who preferred a smooth channel to a rocky one, and by residents and visitors who preferred the "water closet" to the outhouse—that is, by practically everyone who knew the river through work or leisure. Europeans shook off the Congress of Vienna's yoke during the revolutions of 1848. But they left the Rhine Commission intact. To one degree or another, nearly everyone got caught up in the Rhine's faster current. The new Rhine was more productive than the old Rhine, and that was that. The power of steam made the stench of industry tolerable—at least until the day arrived when nobody could see the steamboat for the smoke.

Of the Rhine you shall hear enough by-and-by.
It is verily a "noble river"; much broader
than the Thames at full tide, and rolling along
many feet in depth, with banks quite trim,
at a rate of four or five miles an hour,
without voice, *but full of boiling eddies,*
the most magnificent image of silent power
I have seen; and in fact,
one's first idea of a world-river.

—Thomas Carlyle (1852)

CHAPTER 2

Europe's "World River"

The Rhine is Europe's busiest waterway. As the only river linking the
Alps to the North Sea, it channels the flow of trade through Switzerland, Germany, France, and the Netherlands. The Rhine transports over
200 million metric tons of goods annually, far more cargo than is carried
on any other European waterway. Seabound freight travels so fast and freely
between Rotterdam and Basel that it is almost absurd to think of modern-
day Switzerland as a landlocked country.[1]

The Rhine is puny compared with the Earth's mightiest streams. Its total
length of around 1250 kilometers (775 miles), drainage basin of 185,000
square kilometers (71,400 square miles), and average delta discharge of
2200 cubic meters per second (2875 cubic yards/second) place it ninth
among Eurasian rivers,[2] and it barely ranks in the top hundred rivers worldwide. The Nile is five times longer. The Amazon basin is more than thirty
times greater in breadth. The Danube, Central Europe's largest river and
the Rhine's nearest neighbor, carries nearly three times the water volume.
But what the Rhine lacks in size it compensates for in navigational arteries
and tributaries, including the Lippe, Ruhr, Mosel, Lahn, Nahe, Main,
Neckar, Ill, and Aare. The Rhine is connected to the Baltic via the Rhine-
Herne and other German canals, to France via the Rhine-Marne canal,
to the Mediterranean via the Rhine-Rhône canal, and (since 1992) to the
Black Sea via the Rhine-Main-Danube canal.

The Port of Rotterdam—"The Gateway to Europe" on the Rhine
mouth—is the number one destination of Mideast oil tankers and the
world's largest ocean harbor. Its annual throughput of bulk goods and general cargo (282.4 million metric tons in 1993) edges out even the great Asian
ports of Singapore (273.7 million), Kobe (168.7 million), and Shanghai
(167.9 million).[3] Duisburg-Ruhrort, situated atop Europe's largest under-

ground coal deposits, boasts the world's largest inland harbor. Over seven hundred ships and barges cross the Dutch-German border each day, an average of one vessel every two minutes round the clock.[4] Cologne is Europe's busiest rail hub for passenger and freight trains. Frankfurt's Rhine-Main Airport is Europe's second busiest air traffic center. No other river possesses such an extensive network of navigable tributaries, canals, seaports, inland harbors, and transportation links. Only the Mississippi-Missouri system transports more freight each year.

The Rhine drainage basin is thick with people. Nearly fifty million humans live within its watershed, many of them packed together in just six urban conglomerates directly on its riverbanks: the Dutch "Randstad" region (Rotterdam and its urban environs); Germany's Rhine-Ruhr (Duisburg-Essen-Cologne), Rhine-Main (Frankfurt-Mainz-Wiesbaden), and Rhine-Neckar (Mannheim-Ludwigshafen) regions; France's Strasbourg; and Switzerland's Basel canton.[5] Twenty million people rely on

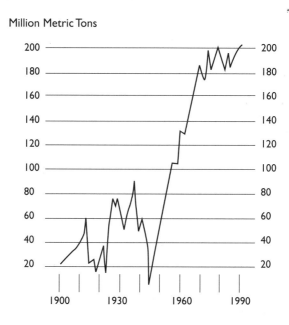

Figure 2.1 Volume of goods transported on the German Rhine, 1900–1990 (Adapted from Deutsche Kommission zur Reinhaltung des Rheins, *Rheinbericht 1990*, 13)

the Rhine for their drinking water (thirty million if Lake Constance is included), and millions of others are indirectly dependent on the river for urban services, sanitation, transportation, and employment. Some of Europe's largest iron, steel, automobile, aluminum, textile, potash, and paper firms are headquartered on the Rhine, as are five of the world's biggest chemical-pharmaceutical firms: Bayer, BASF (Badische Anilin- und Soda-Fabrik), Aventis (formerly Hoechst), Novartis (formerly Ciba-Geigy and Sandoz), and Roche (formerly Hoffmann–La Roche).[6] The textile industries of Voralberg (Austria) and the potash mines of Alsace (France) use the Rhine for waste disposal, while farmers in Baden (Germany) and the Netherlands are dependent on the river for crop irrigation. Without the Rhine, Switzerland's aluminum plants would lack the hydroelectric power needed for production and access to world markets. Also important for many industries are the hundreds of conventional power plants, as well as many nuclear ones (at Beznau, Gösgen-Däniken, Leibstadt, Biblis, Philippsburg, Fessenheim, Cattenom, and Neckar-Westheim) that utilize the Rhine and its tributaries for cooling purposes.

The Rhine, in short, is a classic example of a "multipurpose" (or "multi-objective") waterway, used for transportation, power generation, industrial production, urban sanitation, and agriculture. The river supplies so many diverse human needs that at any given moment as much as *one-sixth* of its water is flowing in and out of pipelines *along its banks*. A single drop of Alpine water might be drawn, utilized, cleaned, and returned to the river-bed many times over on its journey downstream—by hydroelectric dams for energy, by steel firms for production, by chemical companies for waste removal, by power plants for cooling, by textile firms for dyeing, by farmers for irrigation, by mines for coal washing, by paper companies for bleaching, and by cities for sanitation.[7]

Geopolitical Features

The Rhine basin cuts a broad swath across the heart of western Europe as the river moves in a northwesterly direction from the Alps to the North Sea. There are today eight sovereign states that lie partly or wholly within its watershed: Switzerland, Austria, Liechtenstein, Germany, France, Luxembourg, Belgium, and the Netherlands. Of these, four are particularly important to the Rhine's political ecology: Switzerland, where most of the river's headwaters and Alpine stretches can be found; Germany, home to over half (100,000 square kilometers) of the river's drainage basin; France, which shares the expansive Rift valley floodplain with Germany; and the Netherlands, which is all but synonymous with the Rhine delta.[8]

Terms such as "Swiss Rhine," "German Rhine," "French Rhine," and "Dutch Rhine," however, do little justice to the enormity of the river's diverse geomorphology, ecology, and climate. Scientists have therefore inserted several useful divisions and subdivisions into their mental maps of the river (see figure 2.2). Most importantly, the Alpine stretches of the Rhine have hydrological and biological conditions fundamentally different from those on the non-Alpine stretches. The Alpine areas include the "Alpenrhein tributary system," the "Aare tributary system," and the "High Rhine," all of which lie upstream (south) of Basel. Trout and grayling are the characteristic fish species, with alder and willow the most common trees. The non-Alpine stretches include the "Upper Rhine," "Middle Rhine," "Lower Rhine," and "Delta Rhine," all of which lie downstream (north) of Basel.[9] These stretches belong to the barbel, bream, and (at the delta) flounder regions, with oak and elm as the dominant trees—or at least they did until river engineering obliterated these distinctions.

The *Alpenrhein tributary system* is generally regarded as the Rhine's main headwaters. The flow begins in southeastern Switzerland along the southern flank of the St. Gotthard massif, in the canton of Grisons (also known as Graubünden). Two headstreams, the Hinterrhein and Vorderrhein, collect glacial runoff and melting snow from hundreds of tiny rivulets and funnel the water down the narrow crags and gorges to the valley below. The Hinterrhein (57 kilometers long) flows northward from the Paradise Glacier near the Rheinquellhorn (3202 meters high) down the Via Mala, a spectacularly steep and dangerous canyon. Below Thusis, it loses some of its Alpine character and begins to wind its way through the Domleschg valley toward Reichenau. The Vorderrhein (68 kilometers in length) cascades eastward from Lake Toma at the foot of Mount Badus (2928 meters in height) on the southern flank of the Gotthard massif down the steep Bündner Oberland valley. The two headstreams merge to become the Alpenrhein (100 kilometers) at the small Swiss town of Reichenau, just north of Chur. The Alpenrhein then winds its way through a wide valley choked with glacial debris, its riverbed serving at times as a border separating Switzerland, Liechtenstein, and Austria, before flowing into Lake Constance (545 square kilometers in surface area). Biologically the Vorderrhein and Hinterrhein belong to the Trout Region, the Alpenrhein to the Trout/Grayling Region.[10]

For convenience, some geographers pinpoint Lake Toma, high in the

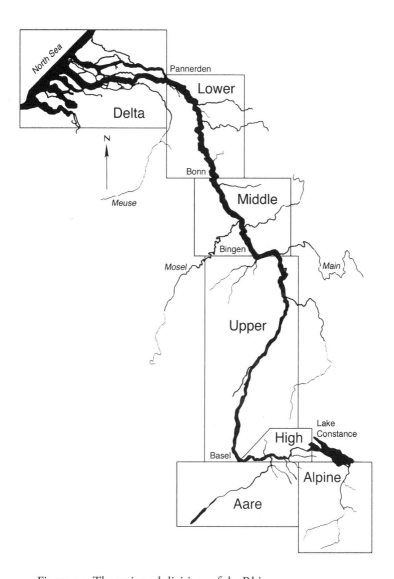

Figure 2.2 The major subdivisions of the Rhine

Grisons Alps, as the principal "source" of the Alpenrhein (and therefore the Rhine), while others choose Grisons' capital city, Chur, as a starting point. But in reality Alpine streams have no single source, and therefore no easily identifiable geographic location, from which they spring. In Rhaeto-Romansch, the local language of Grisons, "Rein" simply means "flowing water" or "river." The various mountain streams are distinguished by the gorges through which they flow: the Rein de Maighels, the Rein de Medel, the Rein de Sumvitg, and so on down the valley ridges. The Alpenrhein is the sum of all these Alpine flows. It is literally "The River."

The *Aare tributary system* originates in the western canton of Aargau, on the opposite side of Switzerland from Grisons. The Aare river (295 kilometers in length) flows northwest through the Brienz, Thun, and Biel lakes (in part due to nineteenth-century engineering work), then flows northeast along the foot of the Jura Mountains. There it picks up the tributary waters of the Reuß and Limmat shortly before it merges into the Rhine at Waldshut, a small town about a hundred kilometers downstream from Lake Constance. The Aare can make a legitimate claim to being the Rhine's pre-eminent Alpine feeder stream. It is longer than the combined length of the Vorderrhein, Hinterrhein, and Alpenrhein. It captures three times as much of the Swiss Alpine watershed and discharges more water into the Rhine at the point of merger (563 cubic meters per second as opposed to 440 for the Rhine).[11] It is also the mother lode of the "Rhine Gold" that once washed downstream to become part of Germanic fables, operas, and jewelry. But for the predilections of the ancient Romans, and the stubbornness of modern geographers, the Aare might well have achieved its rightful pride of place over the Alpenrhein as the Rhine's headwaters. As it stands, however, its watershed is not included in tallies of the Rhine's overall dimensions, and its centrality to the river's drainage patterns is therefore often overlooked.

The river first acquires the name "Rhine" (as distinct from Alpenrhein or Aare) at the westernmost end of Lake Constance, where the water leaves the lake and begins its northwesterly journey to the North Sea. The exact demarcation point is an artifact of the Rhine Commission: kilometer "0" commences at the Constance bridge. Each subsequent kilometer is marked by large signs on the banks until the river reaches Hoek van Holland at kilometer 1033. This level of precision is a bit deceptive. Some stretches of the river have been shortened by engineers since the kilometer signs were posted, while others have elongated themselves as the banks shifted naturally. There are, therefore, numerous "short" and "long" kilometers. Furthermore, sea traffic on the river begins at kilometer 149, at Rheinfelden (just upstream from Basel), and not, as the navigational signposts might suggest, at Constance. To add to the confusion, Lake Constance is treated as a river widening rather than a separate geologic formation, and thus the Alpenrhein tributary system is usually reckoned as part of the Rhine's total length. That is why the Rhine is 1250 kilometers long, and not 1033 kilometers as the signpost at Hoek van Holland might suggest. Engineering maps sometimes even designate Alpenrhein locations as negative kilometers read backward from Lake Constance, though no such signposts exist on the Alpenrhein itself.

The *High Rhine* (from kilometer 0 at Constance to kilometer 168 at Basel) shoots westward between the Jura Mountains and the Black Forest, serving for most of its way as a border between Switzerland and Germany. Alpine qualities still prevail: the current is swift and unpredictable, owing to the influx of many mountain tributaries, chief among them the Aare tributary system at Waldshut. Biologically the High Rhine belongs to the Grayling/Barbel Region. Local navigation on the High Rhine, a hazardous enterprise in the past, has been eased somewhat by the construction of hydroelectric dams and locks, giving the river the profile of a descending staircase. Aluminum, textile, chemical, and other industries have grown up alongside the dams to take advantage of the plentiful water supply and cheap hydroelectricity. In the early twentieth century, these industries began to push the Swiss and German governments to construct shipping lanes along the High Rhine from Basel to Lake Constance capable of accommodating large vessels.[12] Had these plans been realized, the Alpine village of Feldkirch (Austria) might well have become the world's highest "sea" port. No doubt this is what the Rhine Commission had in mind when it chose Lake Constance as the starting point for its navigational signs. However, circumventing the Rhine Falls (30 meters) at Schaffhausen proved too costly, and therefore the High Rhine has never been able to handle anything more than local ships and barges.

The *Upper Rhine* (kilometer 168 at Basel to kilometer 528 at Bingen) is the uppermost stretch of the river open to North Sea traffic. The Rhine leaves the Alps at this point and begins its northward flow, picking up the waters of the Neckar and Main along the way. Politically, the Rhine forms the eastern boundaries of Alsace and Lorraine, the two provinces once claimed by both the Germans and the French. Geographically, the Upper

Rhine flows through the great Rift valley, a broad and natural floodplain at the foot of the Alps, bounded by the Vosges and Haardt/Pfälzerwald ranges to the west and the Black Forest and Odenwald to the east. The Upper Rhine's natural velocity is slow, its natural courseway braided and curved, its natural width constantly in flux. But it has been so thoroughly reengineered that it has lost its original character and is virtually indistinguishable from a canal. It was here that German engineers first began to shorten and straighten the river as well as drain its wetlands, clear its forests, and scoop out its islands. They removed 82 kilometers of oxbows and curves between Basel and Worms alone, reducing the river distance between these two cities by almost 25 percent. It was also here that French engineers in the 1920s designed and built the Grand Canal d'Alsace (from Basel to Breisach), a testimony in cement to France's permanent link to Rhine shipping.

Today, the Upper Rhine has been "tamed" from an engineering as well as a political perspective: it is so tightly harnessed and controlled that it rarely floods; and Germany no longer contests France's control over the provinces of Alsace and Lorraine. Biologically, the Upper Rhine once straddled the Grayling/Barbel Region and the Barbel Region, but river rectification has erased these divisions (as it has all the way downstream to Rotterdam). It also once hosted a broad expanse of now-vanished oak and elm forests. Nonetheless, for all its artificiality and denuded look, the Upper Rhine is still home to much of the river's remaining original flora and fauna. This paradoxical situation stems from the fact that the original riverbed still flows (albeit with only a small amount of water) just east of the Grand Canal, providing the last remnants of a natural river terrain as a small sanctuary for native animals and plants.

The *Middle Rhine* (kilometer 528 at Bingen to kilometer 655 at Bonn) etches its way through the Rhenish Slate Mountains, where it merges with the Mosel, its largest tributary, at Koblenz. Biologically, the Middle Rhine once belonged to the Barbel Region. This stretch of the river—home to the Lorelei, the Bingen reef, Drachenfels, Bacharach, St. Goar, Kaub, and vine-clad hillsides—is where most tourists congregate on their visits to the Rhine. It is also the first stretch of the river basin that is solidly "German" on both banks. Outwardly, the Middle Rhine appears more natural than other parts of the river, largely because its high hills and canyon walls have constrained industrial growth and urban sprawl. But extensive engineering work has been carried out below the surface of the water where it is not visible to the eye. Hundreds of underwater rock hazards have been removed from its bed and bank, including those that once made the Lorelei so dangerous to ships, Heine's siren legend notwithstanding. Also, a large shipping lane has been blasted through the Bingen reef, a quartzite vein that runs across the Rhine bed much like a natural dam. Several sections of the riverbed have been deepened to slow the flow, and the channel narrowed to make space for highways, train tracks, hotels, and restaurants. Bacharach, for instance, built its town wall many centuries ago directly at the river's edge; today that wall is separated from the river by a parking lot, promenade, highway, and rail line.

The *Lower Rhine* (kilometer 655 at Bonn to kilometer 867 at the Pannerden canal just north of Lobith-Tolkamer on the German-Dutch border), still "German" for most of the way, is quite different from the Middle Rhine. Biologically it once belonged to the Bream Region. Unfettered by mountain hindrances, the river is free once again to meander in a northwesterly direction toward the Netherlands. Its trajectory takes it through the Rhineland-Westphalian coal region (the Ruhr in common parlance), where it picks up the waters of the Erft, Wupper, Ruhr, Emscher, and Lippe along the way. Much like the Upper Rhine, it has been shortened, straightened, and canalized to improve navigation. Most of its islands have also been removed. To a far greater extent than anywhere else on the river, the Lower Rhine's entire drainage basin has been massively reengineered to service the needs of industry. It is, in fact, the only stretch of the river where as much attention has been paid to the catchment area as to the riverbed itself. Large dams now capture virtually every drop of runoff from the water-rich hills of the Sauerland and Bergisches Land. These dams provide fresh water for urban and industrial use, and keep the Ruhr running high even during dry spells. The Emscher and Erft have been turned into open-air sewers, while the Lippe has become the main feeder stream for the region's canal network. The Rhine, meanwhile, functions as the ultimate dump, gathering coal residues, petrochemicals, and other industrial wastes and "exporting" them downstream to the Dutch delta. Urban congestion exacerbates the problem, for the Lower Rhine is home to the conurbation of cities that make up the Rhine-Ruhr region, with Essen as its geographic and political center. The Lower Rhine and its tributaries, especially the Erft and Emscher, are among the most polluted river systems of Europe and the focal point of much environmental controversy today.

The *Delta Rhine* (kilometer 867 at the Pannerden canal to kilometer 1033 at Hoek van Holland) flows westward through the Netherlands to the

North Sea. The Pannerden canal—built by the Dutch in the eighteenth century to ease shipping and to control flooding—bifurcates the Rhine into the Waal and Nederrijn-Lek, both of which flow into the common Rhine-Meuse delta. The Waal is the main Rhine branch: it receives two-thirds of the Rhine's waters during high-water periods, three-fourths during low-water periods (annual average: 70 percent). The rest of the flow goes to the Nederrijn-Lek (18 percent) and to a third delta branch, the IJssel river (12 percent). The Delta Rhine is linked to Rotterdam via the Waal and Nieuwe Waterweg, to Amsterdam via the Amsterdam-Rhine canal, to Antwerp via Scheldt-Rhine canal, and to the IJssel estuary (formerly the Zuiderzee) via the IJssel—thus permitting Rhine traffic to reach the North Sea by a wide variety of harbors. Rotterdam is far and away the most important of these harbors. The Delta Rhine is less cluttered industrially than the Lower Rhine, except around Rotterdam, where oil refining and petrochemical production take place on a grand scale. Biologically, it once straddled the Bream Region and the Sea-Bream/Flounder Region.

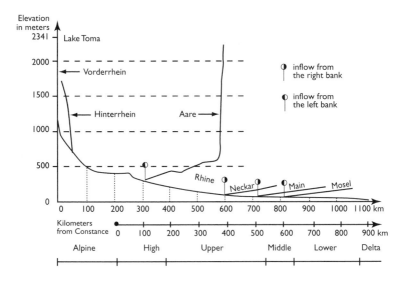

Figure 2.3 Elevation profile of the Rhine from Lake Constance to Rotterdam (Adapted from CHR, *Der Rhein unter Einwirkung*, 15)

Hydrological Features

The Rhine has the longitudinal profile of a typical river: it drops sharply and flows quickly in its Alpine stretches, becomes slower and braided in its middle stretches, and becomes sluggish as it approaches its delta (see figure 2.3). Like most rivers, it picks up more and more water along the way, averaging a discharge of 250 cubic meters per second at the Alpenrhein mouth, 1100 at Basel, 1400 at Worms, 1600 at Mainz, 1700 at Koblenz, 2000 at Andernach, and 2200 at Emmerich.[13]

The Rhine flows more steadily than most rivers, making it well suited for year-round transportation. This steady flow is largely owing to its naturally balanced Alpine and non-Alpine discharge regimes (though engineering projects also play a role). The Alpine tributaries shed their waters mostly during the early summer months when the mountain snowmelt is greatest. Lake Constance acts as a natural reservoir, capturing the Alpenrhein's meltwater and delivering it downstream at a relatively even pace. Seasonal variations are far more noticeable at the Aare's point of merger at Waldshut, a hundred kilometers below Constance, largely because the Aare tributary system lacks any natural or human-made reservoir large enough to capture the bulk of the meltwater before it pours into the High Rhine. The Rhine's non-Alpine tributaries—which can be defined as all feeder streams north (downstream) of Basel—follow exactly the opposite discharge pattern. Fed largely by the rain and snowmelt from French and German mountains (the Vosges, Haardt/Pfälzerwald, Odenwald, Hunsrück, Eifel, Taunus, Westerwald, Bergisches Land, and Sauerland), these tributaries characteristically run high during the cold and rainy winter season and low during the dry summer months (see figures 2.4 and 2.5).[14]

Measured over the course of a year, almost exactly half the Rhine's water comes from the Alps, the other half from the tributaries north of Basel. Severe fluctuations are thereby counterbalanced, allowing for a relatively steady flow year-round into the Dutch delta. In effect, the Alps hold winter precipitation in the form of snow until the summer months, giving the non-Alpine regions a chance to shed their winter accumulation before the Alpine thaw. These seasonal variations can be statistically verified: in winter over 70 percent of the Rhine's waters come from the tributaries north of Basel, while in summer the Alpine streams provide over 70 percent.[15]

"Steady flow" is a relative term for any river: the Rhine's low-to-high flow ratio stands at 1:16 on the Alpine stretches (measured at Basel), and 1:21 on the non-Alpine stretches (measured at the German-Dutch border).[16]

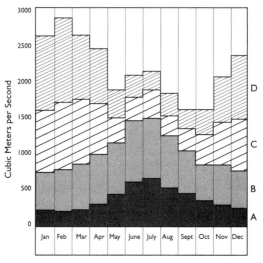

Figure 2.4 The Rhine's annual hydrological regime (Adapted from Ritter, *Le Rhin*, 38)

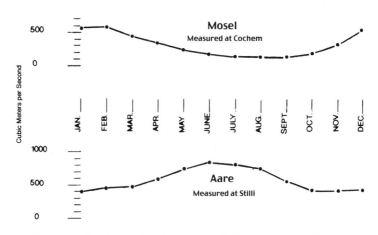

Figure 2.5 Graph showing the opposite discharge regimes of an Alpine (Aare) and non-Alpine (Mosel) Rhine tributary (Source: Ayçoberry, *Une histoire du Rhin*, 322)

Though these variations are small in comparison to many rivers, they are significant enough to translate into plenty of "wet" and "dry" years. Moreover, the foehn—a warm Alpine wind that blows off and on all year from south to north—often contorts the precipitation patterns in Switzerland and Austria. When winds cause an early spring snowmelt, the Alpine tributaries swell faster than usual, causing the Rhine to overspill its banks at the Alpenrhein mouth or at points along the High Rhine. The French and German mountains are just as fickle: long wet winters marked by intermittent thaws will set in motion a tributary swell north of Basel. Spillovers can occur at any number of points along the way. All or parts of the Rhine experienced severe flooding in A.D. 674, 886, 1124, 1342, 1573, 1672, and 1784, the ones of 1342 and 1784 being among the worst in the river's recorded history. Floods occurred in 1824, 1845, 1876, and 1882/83 as the rectification work took place, the first and last of which caused some consternation as to whether the flooding was at least in part engineering related. In the twentieth century, for which the records are most complete, full or partial floods occurred in 1910, 1919/20, 1925/26, 1930, 1939, 1944, 1945, 1947/48, 1955, 1958, 1970, 1983, 1988, 1993, and 1994. Dry periods also occur intermittently, most recently in 1921, 1947, 1949, and 1976.[17] The record peak discharge (measured at the Dutch-German border) occurred in 1926 when the Rhine flow stood at 12,600 cubic meters per second. The record low came in 1947 when the discharge at the Dutch-German border measured a mere 620 cubic meters per second.[18]

"Flood" is a highly anthropocentric term, rooted in the human proclivity to think of a river as having a fixed length but no prescribed breadth, with the result that the floodplain is often used for farms and settlements as if it were not part of the river system. In actuality, water simply follows the path of least resistance from elevated areas to sea level, using as much of the landscape as necessary at any given time. Depending on weather patterns, a river might be flowing high or low, sometimes utilizing its floodplain to carry water, sometimes not. During low-water periods, a river typically flows in one channel or along narrow channel braids. During high-water conditions it utilizes its floodplain more fully, allowing its true breadth to become visible to the eye. When humans are there to witness these high-water flows, and especially when their lives and properties are affected, they record a "flood."

How and where these overspills occur depend on the morphological, topographical, and precipitational conditions of a river at any given time

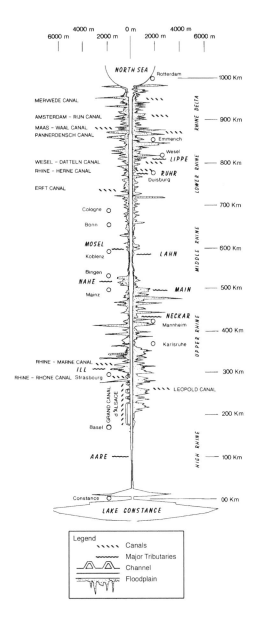

Figure 2.6 Schematic diagram of the Rhine and its flood zones
today (Source: CHR, *Das Rheingebiet*, C.13.1)

(see figure 2.6). During high-water periods, rivers absorb the extra water much like a python digests its prey: a bulge (or swell) appears as the water passes downstream. In broad river basins such as the Rhine, with many tributaries and subtributaries, the trunk river tends to experience several intermittent swells in addition to the main swell. This is because smaller tributaries tend to crest and recede faster than larger ones, and larger tributaries faster than the trunk river itself. Timing is everything. The flood danger is lowest if the tributaries have a chance to recede before the main trunk crests. It is highest if the large tributaries and main trunk reach their peak discharges at the same time in the same place.

Nineteenth-century river engineering considerably altered the Rhine's natural hydrology and therefore its overspill pattern. Under natural conditions, the Upper Rhine is the stretch most susceptible to spring snowmelt for an obvious reason: it is situated in the broad Rift valley directly at the base of the Alps. The lower stretches of the river—the Middle and Lower Rhine—were relatively shielded from the ravages of Alpine runoff because the Rift valley absorbed much of the floodwater like a sponge, greatly reducing the amount of water that poured downstream. (The Romans, for instance, found it prudent to site Strasbourg and Freiburg on high ground away from the mercurial Upper Rhine, but placed Mainz, Koblenz, Bonn, and Cologne almost directly on the banks of the Middle and Lower Rhine.) Because the Rift valley absorbed and temporarily stored much of the Alpine runoff, the Rhine's non-Alpine tributaries (the Main, Mosel, Ruhr, and so on) generally had sufficient time to dump their floodwaters into the trunk river before Alpine floodwaters arrived downstream. Intermittent swells might cause the Rhine's lower stretches to flow high for several days or weeks, but massive overspills were often avoided except under extreme conditions—the so-called hundred-year floods, which were as devastating as they were rare.

In the nineteenth century, however, engineers began shortening the Upper Rhine by 82 kilometers, which greatly reduced the time it took for floodwaters to travel downstream. They also constructed artificially high banks to keep the water constricted in one bed. Excess water, which in the past would have saturated the Rift valley's water table, was now funneled down the narrow gorges of the Middle Rhine and into the Lower Rhine floodplain. The river's current became swifter and its route more direct: the travel time for water swells from Basel to Karlsruhe alone was reduced by two days, from 65 hours before the rectification work to 24 hours today.[19] As a consequence, water from the Upper Rhine now reaches the Middle and Lower Rhine more quickly, sometimes cresting before the tributaries have receded. The result is flooding that can be attributed as much to human engineering as to acts of nature. Flood dangers are highest in years when long rainy winters are interrupted by unseasonable thaws, putting a maximum burden on the entire Rhine catchment area simultaneously. Inundations struck the Middle and Lower Rhine in March 1983, again in March 1988, again in December 1993, and again in December 1994—four hundred-year floods in the span of twelve years. Engineers have, in effect, partly shifted the flood burden from the base of the Alps to the Middle and Lower Rhine stretches.[20]

The Rhine Commission

The Rhine Commission was established by the Congress of Vienna in 1815 and formally constituted a year later.[21] Far from being an arcane and peripheral feature of the negotiations, the Rhine Commission was championed by the Congress's highest ranking statesmen, including the Prussian diplomat Wilhelm von Humboldt (brother of the famous naturalist Alexander von Humboldt) and the Austrian Foreign Minister Clemens von Metternich, who was born and raised in the Rhine city of Koblenz. The Rhine was considered important enough to warrant a special subcommittee devoted solely to river affairs. Composed mostly of delegates from the Rhine states, the subcommittee wrangled for months before it decided to create the Rhine Commission as an umbrella organization with far-reaching powers over the river's economic affairs. Once established, however, it proved popular: imitator commissions were soon in place on the Elbe (1821), Weser (1823), Ems (1853), and Danube (1856).

Hardly was the ink dry on the Vienna accords when a second event rocked the Rhine world: the arrival in 1816 of the British steamship *Prince of Orange* in Cologne, belching so much smoke and soot that it looked like a moving volcano to those who witnessed its arrival.[22] Before the steam age, bulk trade on the Rhine consisted mostly of a *downstream* flow of raw timber (oak, fir, spruce, pine) from the Black Forest to the Netherlands, the so-called Holland rafts, in which the vessel and the product were one and the same thing.[23] *Upstream* trade was an onerous undertaking at best, requiring towpaths and animal or human muscle, and the need to change ships frequently along the way. A journey could take weeks or more depending on the vagaries of the wind, the speed of the current, and the availability of horses and men. The *Prince of Orange* heralded the day when it would be feasible to haul bulk goods (coal, ores, grains) quickly and cheaply upstream by steam-powered tugs and barges, greatly increasing the number of industrial ports, commercial centers, and trading routes on the upper stretches of the river. It also, of course, signaled the era of smoke-filled skies and industrial air pollution along its banks.

Shortly thereafter, in 1817, a third novelty hit the Rhine: the onset of the Tulla Project, designed to eliminate the river's oxbows, braids, and islands. Initially restricted to flood-control measures on the Upper Rhine, where the river meandered through the Rift valley floodplain, the work was later extended to other parts of the river, and by century's end the entire river, from the Swiss Alps to the Dutch mudbanks, had been thoroughly reengineered. Without major rectification work, the Rhine would never have been able to provide the regular water flow needed for irrigation, urban services, and year-round transportation, or to keep pace with the ever-increasing size of the new ships; and without extensive dredging, Rotterdam could never have been turned into a world-class harbor.

The period 1815 to 1817 thus serves as a convenient "birth date" for the modern Rhine, and one that highlights the international scope of the undertaking. But historical events have a tendency to overspill their narrative channels, greatly muddying the chronological flow. The actual roots of the modern Rhine—and indeed of all European rivers—reach back to the Italian Renaissance, and especially to the time span between the publication of Benedetto Castelli's *Della misura dell'acque correnti* (1628) and Domenico Guglielmini's *Della natura de' fiumi* (1719). A pupil of Galileo, Castelli inaugurated the study of modern river hydraulics ("theory of rivers," "hydrodynamics," and "fluid resistance"). He also mathematized river engineering by calculating for the first time the formula needed to determine the amount of water flowing in any given river. Guglielmini, meanwhile, provided subsequent generations of engineers with the first practical guide for taming and controlling rivers. This guide was based on two insights. First, stagnant waters in pools and swamps around rivers were associated with the spread of certain diseases, notably typhoid and malaria; therefore, river rectification projects should concentrate on channeling water from its source to its mouth swiftly and efficiently enough to prevent backwaters and pools. Second, a river with a broad or deep main channel tends to flood less often than a river with a narrow, shallow, or braided channel; therefore, in the interest of maximizing agricultural production and minimizing overspills, rivers should be harnessed into one main channel. Preferring straight to curved and swift to slow, Italian engineers worked out the mathematical formulas that rendered it feasible for the first time to widen and shorten rivers on a scientific basis.[24] Johann Gottfried Tulla's famous maxim—"As a rule, no river or stream needs more than one bed"—was a condensation of two hundred years of Renaissance river theory.

Italian ideas quickly spread to Switzerland, the Netherlands, the German states, and above all to France, where they became part of the "enlightened" curricula of the École des Ponts et Chaussées, the École Polytechnique, and other military-engineering schools in the eighteenth century. The important new dimension these schools added to the Italian tradition was the notion that river engineering was central to the state-building process—that an "enlightened" government could reclaim land for agriculture, manage its water resources, and improve commerce at one and the same time. "Each kingdom, each province, each city has its hydraulic needs," wrote Pierre Louis Georges Du Baut in the great engineering classic of his time, *Principes d'hydraulique* (1779): "necessity, convenience, or luxury cannot do without the help of water; water must be brought into the very center of our dwellings; we must protect ourselves from its ravages, and make it work the machines that will ease our discomforts, decorate our dwellings, embellish and clean up our cities, increase or preserve our holdings, transport from province to province, or from one end of the earth to the other, everything that need, refinement, or luxury have made precious to us; large rivers must be contained; the beds of smaller rivers must be changed; we must dig canals and build aqueducts."[25]

As a Paris-trained student of hydraulics, Tulla was fully schooled in the Enlightenment tradition. "In agrarian regions," he wrote in his Rhine blueprint, "brooks, streams, and rivers should, as a rule, be canalized, and their flow harnessed to the needs of people who live along their banks."[26] He also fully shared in the optimistic spirit of his era. "Everything along this stream will improve once we undertake the rectification work," he claimed in his Rhine blueprint. "The attitude and productivity of the riverbank inhabitants will improve in proportion to the amount of protection their houses, possessions, and harvests receive. The climate along the Rhine will become more pleasant and the air cleaner, and there will be less fog, because the water table will be lowered by nearly one-third and swamps will disappear."[27]

Common to nearly all European engineering textbooks was the notion that rivers were potentially the "enemies" of humans and therefore in need of being "domesticated," "tamed," or "harnessed"—three of the most commonly used metaphors for river manipulation at the time. The fact that rivers posed serious limitations to the movement of troops during wartime further fed this militant approach to water engineering: armies needed engineers skilled in the art of building bridges and overcoming other hurdles imposed by water. (Since the same engineers were often in charge of constructing both fortresses and flood works, it should come as no surprise that riverbank fortification came to resemble fortress walls.) Tulla, himself a military engineer, called his Rhine blueprint a "general operational plan" for a "defense against a [Rhine] attack" in 1812. In 1825, a government deputy praised the Tulla Project with the words "finally we have a war strategy against the Rhine's waters that is intelligently conceived and therefore has a chance of succeeding." And when the last and most difficult phase of the Tulla Project was completed at Istein in 1876, the Rhine engineer Max Honsell triumphantly proclaimed: "the correction has succeeded here too, but only after a stubborn and protracted battle with the river."[28]

Also common to the language of river engineering was the notion that free-flowing rivers were by nature somehow imperfect or defective, and therefore in need of improvement—here "rectification" and "amelioration" were the most common terms. Blueprints for canal construction commingled with blueprints for river rectification, and with the blueprints went the mentality of improvement. Widespread among European engineers was the perception that the perfect or "ideal" river was really a canal: straight, predictable, easily controlled, specifically designed for navigation, not prone to flooding, easily contained within a single channel, and not so sluggish as to breed disease. It is only a slight exaggeration to say that all of Europe's rectified rivers were designed and built to resemble canals; and that nearly all of the projects stemmed from the Zeitgeist of the Enlightenment, the belief in the perfectibility of humans and rivers.

If river rectification had its intellectual roots in the Italian Renaissance, the international river idea had an indisputably French genealogy. When French revolutionary troops poured across the Rhine in 1792, they swept away forever the political structure that had prevailed on the river since medieval times. After losing the Austrian Netherlands (Belgium) to France in 1795, the Habsburgs never returned to the Rhine as a riparian power. The *ancien régime* of lilliputian states, knightly estates, and ecclesiastical principalities gave way to new middle-sized modern Napoleonic states—Baden, Württemberg, Hesse, Nassau, Berg, Westphalia—as well as political unity under the aegis of the French-controlled Confédération du Rhin (1806–13). France's direct influence ended with Napoleon's defeat in 1814, but the Vienna delegates decided against restoring the empty shell of the Holy Roman Empire or the microgovernments of yesteryear. Instead, they accepted Baden, Hesse, and the other Napoleonic creations as legitimate, thus ensuring the existence of riparian states large enough to engage in rectification work on a grand scale. To have done otherwise would have meant the restoration of ninety-seven formerly independent German political entities between Alsace and the Netherlands—a return, in other words, to the landscape of petty quarrels that had previously made political cooperation so difficult.

On economic issues, too, the French Revolution's legacy was immense, especially as regards the establishment of liberal trade policies. The Vienna diplomats found it useful to rely on the Convention de l'Octroi (signed between Napoleonic France and its right-bank German satellite states in 1804) as a guide in formulating their own unified commercial codes.[29] Similarly, the Magistrat du Rhin, a Napoleonic agency created in 1808 to handle France's commercial and engineering matters on the Rhine, served as a model for the Rhine Commission (Tulla first presented his Rhine blueprint while serving as Baden's liaison officer to the Magistrat). The Vienna diplomats, in fact, accepted many Rhine issues as *faits accomplis* more readily than they cared to admit publicly. The very act of agreeing to internationalize the Rhine signaled a tacit acceptance of a major French revolutionary goal: "The restrictions and obstacles formerly placed on Rhine sailing and trading directly conflict with natural law, which all Frenchmen are sworn to uphold," stated the French Executive Council decree of 16 November 1792. "The flow of rivers is a common asset, not given to transfer or sale, of all states whose waters feed them."[30] Though the Vienna diplomats avoided any reference to natural law or Frenchmen, their intent was identical though their wording was more prosaic: "Navigation on the Rhine, from the point where it becomes navigable to the sea [*jusqu'à la mer*] and vice versa, shall be free, in that it cannot be prohibited to any one."[31]

Just as there were many preludes to 1815, so too were there many postscripts. The international river idea existed more happily in the minds of French revolutionaries and Vienna diplomats than it did in those of state and port authorities, where the tendency to view commerce as a zero-sum game lingered long after the spirit of cooperation had dissipated. The Dutch in particular were less than thrilled by the prospect of open and free trade on the Rhine. Ever watchful of their monopoly over colonial tea, salt, and spice imports (their so-called transit privileges), they construed the phrase *jusqu'à la mer* to mean "up to the sea" and not "into the sea," thus exempting their seaports from the terms of the Vienna accords. Similarly, they interpreted the phrase *jusqu'à son embouchure* ("to its mouth") to mean the mouth of the Nederrijn-Lek, the Rhine's secondary delta channel. This interpretation allowed them to exclude the mouths of the Zuiderzee, Scheldt, and Waal (the Rhine's main delta branch), where almost all Dutch trading took place.[32]

Dutch diplomats let these phrases slip into the accords and then exploited them to their advantage. But in practice all that the ambiguity did was provoke reprisals and countermeasures that slowed the momentum toward a more liberal trade regime. The city of Cologne, for instance, refused to give up its own lucrative staple and transfer privileges until the Dutch returned to the negotiating table. Other localities also found ways to keep their prerogatives intact as long as possible, citing the Dutch precedent. It took sixteen years to resolve most of these conflicts (the Mainz Acts of 1831) and another thirty-seven years to resolve the rest (the Mannheim Acts of 1868)—a time lag of fifty-three years, long even by the molasses-like diplomatic standards of the day. It probably would have taken much longer, but in the 1850s cutthroat competition from new railroad lines converted shippers and riparian governments into unabashed free traders.[33]

Prussian aggrandizement also played havoc with the international river idea. The Rhine Commission's charter members included all seven of the then-existing riparian states north of the Alps: Prussia (Rhineland and Westphalia), Hesse, Nassau, Baden, Bavaria (Palatinate), France (Alsace and Lorraine), and the Netherlands. Prussia annexed Nassau after the Austro-Prussian war of 1866, reducing the number to six. Then Baden, Hesse, and Bavaria joined the Bismarckian empire in 1871, as did Alsace and Lorraine, which France lost as a result of the Franco-Prussian war. That meant that from 1871 to 1918 only the presence of the Netherlands on the commission kept the navigable stretch of the Rhine from becoming "Germany's river" as Arndt had proclaimed it to be back in 1813.

France returned to the Rhine with a vengeance at the end of World War I, reclaiming Alsace and Lorraine, moving the commission's headquarters to Strasbourg, and expanding its membership to include Switzerland, Belgium, Italy, and Great Britain, all for the purpose of diluting German influence (Articles 354-362 of the Versailles Treaty). Nazi Germany and Fascist Italy quit the commission in the mid-1930s, and then it all but disappeared when Hitler brought the whole of the navigable Rhine (this time including the Dutch delta) under German control during World War II. It was not until the Federal Republic of Germany rejoined the commission in 1950 that a wholly cooperative spirit prevailed, and the Rhine became a fully functioning international river in spirit as well as in fact.[34]

Overarching authority over the Rhine Commission still rests in the hands of Europe's Great Powers—today under the aegis of the European Union—just as it did in 1815. This control has almost always existed more in principle than in practice. Only once since the Congress of Vienna have Rhine navigational politics reached the highest echelons of European diplomacy: in 1919, at the Paris Peace Conference, when France returned as a riparian power. That was, of course, a significant intervention, and the Versailles Treaty proved to be an explosive issue in its own right. But it was a singular interlude in the long history of the Rhine Commission. Under normal circumstances, diplomatic initiatives have been left to the discretion of the riparian governments and the commission members themselves.

The Rhine Commission met for the first time in 1816 and has met on an annual basis since 1831. It did not meet regularly before 1831, in part because the Vienna statesmen did not establish clear lines of bureaucratic authority and in part because the Dutch transit issue had to be settled before the commission could function effectively. Originally headquartered in Mainz, it moved to Mannheim in 1860, and then to Strasbourg in 1919, each change of location reflecting an upstream extension of the Rhine's navigability. Since it was created as an advisory board, not a policy-making institution with legal teeth, all its undertakings rest on the general consent of its member states. Most issues that come before the commission are of common concern, so its recommendations are rarely ignored, rejected, or thwarted. The commission took the lead in standardizing navigational regulations, police ordinances, and emergency procedures on the river. It oversaw the difficult transition from the age of rafts and sailboats to the age of steamers, diesels, and push-tows. It initiated regulations on the transport of hazardous materials, as well as for the safe removal of ship waste and bilge oil. Almost all blueprints—from a simple bridge construction to elaborate engineering projects—pass through the commission's offices before work commences. The Rhine Commission also undertakes a river inspection once every ten years to assess the impact of previous projects and recommend new ones.[35]

The Congress of Vienna left considerable power in the hands of the riparian states themselves. Issues that affect the makeup and goals of the Rhine Commission are usually handled through the diplomatic channels of the various foreign offices. Ratification of the Mainz Acts (1831), for instance, was possible only after Prussia and the Netherlands resolved their long-standing differences over the scope of the Vienna accords; and the Mannheim Acts (1868) had to wait until all the riparian powers were comfortable with the idea of free trade. Rectification projects that cut across borders invariably required discussions among statesmen from the affected states, and much of this discussion has traditionally taken place within the corridors of the foreign offices rather than the Rhine Commission. The Vienna accords also placed power of the purse at the national level, though costs have often been shared by the regions, cities, and industries that gained the most from the improvements. It was fortunate, for instance, that the economically powerful Prussian state controlled the Middle Rhine, where the costs of removing the underground rocks and barriers were exorbitant. It was equally lucky that the Dutch, with their far-flung commercial and colonial interests, were in charge of expanding the delta ports, and that the Swiss were in the forefront of Alpine diversion and electrical generation projects. Sometimes, however, state politics interfered with the smooth flow of commerce. For decades, the small state of Hesse delayed dredging a navigational channel in the Rheingau (a lake-like river broadening upstream from Bingen) due to lack of funds. The state of Baden, meanwhile, stymied navigational improvements between Mannheim and Strasbourg so that it could preserve Mannheim's status as the Rhine's uppermost port and ensure the profitability of the Baden railway network. And from 1919 to 1976, French and German engineers repeatedly clashed over the construction of the Grand Canal d'Alsace.[36]

At the lowest level of administration come the various national and state river agencies: Baden's Water and Road Construction Department, Prussia's Rhine River Engineering Administration, and the Netherlands' Department of Public Works and Water Management, to mention only a few of the earliest ones. They operate under the jurisdiction of the governments that pay them. It is these agencies that undertake the hydraulic studies, develop the blueprints, oversee the rectification work, and monitor the results. The fact that they are organized along national lines, and function as arms of the state authorities, reinforces the visibility of the individual riparian states in Rhine affairs. But in practice a sense of camaraderie has always prevailed among the engineers who work in these agencies. For the most part, they have all been trained in a handful of technical schools with nearly identical curricula and they all speak the same language of mathematics. Ideas, personnel, and blueprints have been crisscrossing borders from the outset, ensuring a high degree of international cooperation and consensus building, except during times of armed conflict.

To a large degree, therefore, the Congress of Vienna turned over control of the Rhine not so much to the riparian governments as to a group of like-minded Eurocrats and engineers, and gave them carte blanche to redesign the river as they saw fit. The Rhine Commission's influence "from above" in setting overall goals, and the Rhine engineers' influence "from below" over the details, helped to keep the power of the riparian governments in check even in an age of increasingly militant nationalism. The bureaucratic layering was an awkward arrangement at best. But it worked tolerably well because everyone involved shared a common goal: transforming the river into a world-class shipping canal.

In his recent book, *Seeing Like a State*, James C. Scott has argued that many state-sponsored projects designed to improve the human condition go awry because those who conceive and implement them woefully underestimate the complexity of the undertakings upon which they are embarking. Eighteenth-century Prussia set out to improve its woodlands through extensive management, but ended up with ecologically impoverished forests. In the twentieth century, the Soviet Union and China collectivized their farms in order to augment agricultural production, only to reap a harvest of sorrow instead. When Brazil moved its capital from Rio de Janeiro to Brasília in 1960, it abandoned an old and vibrant city for a planned and sterile one. Similarly quixotic was Tanzania's attempt in the early 1970s to move its nomadic population into sedentary villages. "State simplifications" is the term Scott used to account for the failures: Prussian authorities spoke of forests but what they had in mind were annual timber yields; Soviet and Chinese planners counted bushels of grain and rice, but left farmers and peasants out of their equations; Brazilian and Tanzanian leaders "saw" their countries full of cities and villages, but not full of people.[37]

Scott focused mostly on major catastrophes wrought by individual states. But his approach can be adapted to almost any large-scale undertaking, even those involving many states and those that successfully achieve their goals. The Rhine riparian states endorsed "simplifications" on a grand scale: their bureaucrats used the term river when they really meant channel and bed; their engineers fretted over the loss of property from flooding, but paid scant attention to the floodplain; their industrialists and urban planners concerned themselves with the quality of the water that flowed into their factories and cities, but not with the quality of the water that poured out of their sewage drains. These were not states that "saw" the river through a poet's eye, or who "knew" the river in the same way that a fisherman did. For them, the Rhine was a navigational channel and a water source, not a biological habitat. Of the nearly seven hundred treaties, agreements, blueprints, and disputes that came under the purview of the Rhine Commission between 1816 and 1916, not a single one concerned itself with water quality, the floodplain, or biodiversity. In fact, only a dozen even addressed environmental issues at all, and then only obliquely—the transport of arsenic, petroleum substances, and other hazardous materials by ship and barge (but not the influx of those toxins into the water itself).[38]

This singlemindedness of purpose greatly helped the Rhine Commission to achieve its goals expeditiously. Indeed, the commission could never have commenced its work had it been forced to preserve all features of the old river as it created the new one. But because the Rhine Commission viewed navigation, flood control, clean water, and biodiversity as "competing" rather than "complementary" goals, it embedded certain characteristic "failures" into its otherwise successful endeavor. The river that existed in the minds of Eurocrats had no biological life. Not surprisingly, therefore, the river they constructed had little room left for fish and birds. A river, as they defined it, had geographic length but no geographic breadth. They thus constructed a river bereft of floodplain. River water, as they saw it, was a "free commodity." They thus stood passively by as water politics on the new river turned into a free-for-all.

Conspicuous by their absence for more than a century and a half were any significant international, national, or local agencies whose purpose it was to protect (let alone "improve" or "ameliorate") the Rhine as a biological habitat. It would not be until after World War II that mentalities began to favor a more ecologically balanced approach to river management. But by the time the first intergovernmental environmental commissions came into being, the Rhine Commission would largely be finished with its task, and Father Rhine would be encased in a bed of cement and stone.

A hydroelectric plant is not built into
the Rhine river as was the old wooden bridge that
joined bank with bank for hundreds of years.
Rather the river is dammed up into the power plant.
What the river is now—a water power supplier—
derives from the essence of the power station.
In order even remotely to understand
the monstrousness that reigns here, one must ponder
for a moment the difference in meaning between
the phrase "The Rhine" as dammed up into the power
works, and "The Rhine" as uttered out of the art work,
in Hölderlin's hymn by that name.

— Martin Heidegger (1949)

But where is the man
Who can remain free
His whole life long, alone
Doing his heart's desire,
Like the Rhine, so fortunate
To have been born from
Propitious heights and sacred womb?

— Friedrich Hölderlin (1801)

CHAPTER 3

Water Sorcery

The nineteenth and twentieth centuries marked the high point of Rhine engineering. But some modest modifications did occur well before then, almost always when a powerful imperial lord, or a coterie of ambitious local rulers, ruled on its banks.

The Romans boldly spoke of bullying the Rhine—"coercendo Rheno" in Tacitus's memorable phrase.[1] In actuality, they just tweaked the river here and there, constructing a few short-lived bridges and dams, mostly in the delta region, where the river's mercurial braids interfered with the movement of their troops. For centuries they relied on the river to transport wine, salt, fur, marble, and amber, but they never realized their dream of linking the Rhine to the Mediterranean by way of a Mosel-Saône-Rhône canal. Charlemagne, the great Frankish king, thought in similarly grandiose terms. After bringing most of the river and its watershed under his control by A.D. 800, he tried to build a canal to link the Rhine and the Danube via the Main river. But all that his workers managed to construct was a useless ditch, and as soon as his grandsons divided the imperial inheritance the Rhine returned to its more familiar role as a fluid boundary between contested terrains.

The Holy Roman Empire also brought some semblance of unity over the Rhine. Yet always lurking behind the empire's unified facade were the innumerable little states that made a journey down the river a web of border crossings. During the medieval period, the Hanseatic League also established a presence on the Rhine, which acted as a stimulus to trade (mostly in wine and herring) between Cologne and the Baltic Sea. There was even enough trans-Alpine trade in the early modern era for Fernand Braudel to call the Rhine "an arm of the Mediterranean Sea."[2] But because the river remained noted more for its transportation chokepoints—toll booths, cliffs, and reefs—than for its transportation links, it languished as a commercial backwater, at least by Baltic and Mediterranean standards.[3]

Modest navigational improvements did occur slowly over time at a few key locations, almost always where an ambitious local prince or warlord held sway. Oxbows and curves were removed at Liedolsheim, Germersheim, Neupotz, Jockgrim, Kembs, Daxland, Dettenheim, and elsewhere during the medieval era. A nine-meter gap, known as the Binger Loch, was painstakingly carved into the Bingen reef (the quartzite vein that cuts across the Middle Rhine channel) to ease shipping during low-water periods. Small dams were emplaced here and there to control local flooding, and towpaths were built to pull boats upstream along commercially lucrative river stretches.[4] In 1707, the United Provinces of the Netherlands constructed the Pannerden canal, which gave the Dutch the ability to regulate the distribution of Rhine waters at the point where the Waal and Nederrijn-Lek bifurcate.[5] A remarkable feat of eighteenth-century engineering, it was a harbinger of things to come. Not long thereafter, Prussia under Frederick the Great constructed a series of dams on the Ruhr tributary to make it serviceable for coal transportation—yet another feat of engineering that bespoke the future.[6]

It was, however, far more characteristic of Rhine inhabitants simply to accommodate themselves to the rhythms and whims of the river. In Roman times, the town of Breisach lay on the left bank of the Upper Rhine, but by the tenth century the river's banks had shifted to such an extent that Breisach was situated on an island. By the thirteenth century it lay on the left bank again, and after the fourteenth century it was once again on the right bank.[7] Such were the vagaries of the natural flowing Rhine until Tulla began to reengineer it.

Tulla and the Upper Rhine

Tulla was for the Rhine what Napoleon was for Europe: the remaker of worlds and the redrawer of maps. Though later engineers would have a greater impact on the river's profile, no one is more closely associated with the creation of the modern Rhine than Tulla. Born on the right bank, in Karlsruhe, he devoted his entire life to shortening and straightening the Upper Rhine stretch from Basel to Bingen. In his youth, Tulla showed a gift for mathematics and a bent for engineering. Under the patronage of the local prince, the Margrave Karl Friedrich, he studied and traveled in Saxony, Holland, and elsewhere. He rounded out his education from 1801 to 1803 with a stay in Paris, where he read the works of the great Italian and French engineers and studied the latest developments in the science of hydraulics. In 1804, he declined a professorship in mathematics at the University of Heidelberg, choosing instead to become Baden's Chief Water Construction Engineer (after 1817, Director of Water and Road Construction). Always more practical than theoretical, Tulla founded an engineering school in 1807 to prepare students for government service, which later became part of the Technical University of Karlsruhe.[8]

The state of Baden, Tulla's employer, was a Rhine parvenu, one of the few German-speaking states that prospered rather than perished during the French Revolution. In 1770, the year Tulla was born, "Baden" still consisted of two separate margraviates, Baden-Durlach and Baden-Baden, each with many noncontiguous territories on both Rhine banks. After Baden-Baden's ruler died without a male heir in 1771, Karl Friedrich of Baden-Durlach scooped up his territories. Even united, Baden consisted of less than 3600 square kilometers (1400 square miles) of scattered lands. In the first wave of French Revolutionary wars (1792–95), Karl Friedrich sided with the Habsburg forces, as did most riparian rulers, and promptly lost all of his left-bank possessions. But Baden's fortunes turned for the better in 1803 when he received the bishopric of Constance and several other right-bank territories. In gratitude, he joined forces with Napoleon in 1805, an alliance that brought him additional right-bank territories at the expense of his neighbors. By the time of his death in 1811, Baden had quadrupled in size (to 15,000 square kilometers, or 5800 square miles) and enhanced its status to that of a Grand Duchy, having digested some twenty-seven formerly sovereign German entities and a few knightly and ecclesiastical estates as well. Karl Friedrich's grandson and heir, the Grand Duke Charles, married one of Napoleon's adopted daughters. He adhered to the Napoleonic alliance until France's defeat at the Battle of Leipzig in 1813, at which time he read the handwriting on the wall and promptly switched sides. The Congress of Vienna subsequently left the Grand Duchy of Baden intact as a buffer state against renewed French aggression.

Baden's northern neighbor, Hesse (known as Hesse-Darmstadt until 1866), had a nearly identical microhistory. The local landgrave, Louis X, lost his left-bank territories during the French Revolution, but was later compensated with right-bank land at the expense of his neighbors. He sided with Napoleon, became a Grand Duke, then joined the anti-Napoleonic coalition in the nick of time. Territorial adjustments in 1815 (losses on the right bank and gains on the left) turned Hesse into one of the few Rhine states that still straddled the river and consisted of noncontiguous territories (totaling 7800 square kilometers, or 3000 square miles). Aside from Hesse to the north, Baden had two western neighbors: the Bavarian Palatinate (*Pfalz*), a left-bank state (6000 square kilometers, or 2300 square miles) which disappeared during the French Revolution and then was partly restored by the Vienna accords; and Alsace-Lorraine (14,500 square kilometers, or 5600 square miles), the only Rhine region still under French control after 1815.

Rounding out the list of new riparian states were the right-bank Rhineland Province (27,000 square kilometers, or 10,400 square miles) and the left-bank Kingdom of Westphalia (20,200 square kilometers, or 7800 square miles). As elsewhere, these provinces consisted of territories that France had seized, consolidated, and lost between the years 1792 and 1815. Both went to Prussia in 1815, making Prussia from then until 1945 the largest German landholder on the Rhine. Rhineland and Westphalia were situated on the Middle and Lower Rhine, north of Hesse, so they did not fall within the immediate purview of the Tulla Project. But the Prussian government, concerned about the danger of downstream flooding, did initiate a short-lived legal attempt to stop the work in Baden between 1825 and 1832.

Most Baden residents felt little sense of political allegiance to Karl Friedrich or his grandson; they had simply awakened one morning to discover that they were now "subjects" or "citizens" of the new Grand Duchy. Accustomed as they were to their local loyalties (to Heidelberg, Mannheim, and the like), they had every reason to view the Grand Duke in the same way they viewed Napoleon—as a "foreign" usurper with whom one might choose to collaborate but not necessarily embrace. The only thing that all Badeners had in common, and the only thing that the Grand Duchy shared with its neighboring states, was that they all lived along the same stretch of the Rhine. As Director of Water and Road Construction for the Grand Duchy, Tulla was therefore in a position to build the bridges—literally and metaphorically—that would overcome parochial loyalties and instill a spirit of cooperation between Baden and the other riparian states. Diplomacy, state-building, and Rhine engineering were inextricably interwoven.

The Upper Rhine, at the base of the Alps, flows northward from Basel to Bingen through the expansive Rift valley. Before the Tulla era, it was the wildest stretch of the Rhine, with a profile that reflected the classic characteristics of a floodplain: soft and ever-changing banks, a meandering and braided bed, a channel choked with shallows, sandbars, and gravel pits, and pronounced erosion patterns. Its numerous islands and wetlands provided an ideal habitat for spawning fish and nesting birds. But its high water table and stagnant pools also allowed many waterborne diseases (typhoid, cholera, dysentery, malaria) to flourish, rendering the region hostile to human habitation. Frequent floods, moreover, made agricultural production and livestock raising hazardous enterprises (records indicate over fifty major floods between 1290 and 1938).[9] The great flood of 1784, the last major one before rectification work began in 1817, caused as much damage to the crops and villages of the Upper Rhine in a single month as did the French revolutionary troops a few years later.

Tulla spent his first years as Baden's water engineer taking measurements of the Upper Rhine's total length and width, and assessing which oxbows and loops to eliminate. A tireless champion of a common measuring system, he introduced the metric system, a French Revolutionary invention, in Baden. He was the first to utilize the Woltman water gauge to measure the Rhine's velocity and the first to test the river's depth by placing water pegs at specific points, a practice still used on the Rhine. Tulla reckoned that only three Baden cities and sixty-three villages were completely safe from flooding. Another three cities and ten villages lay partly in flood-endangered regions, while two cities and twenty-seven villages lay wholly within the Rhine's natural floodplain. His chief task, therefore, was to manipulate the Rhine channel so that it safeguarded the threatened regions without inadvertently inundating the safe areas.[10]

The Tulla Rectification Project (1817–76) — as his comprehensive flood-control scheme came to be called — provided the Upper Rhine with an artificial bed of uniform width: 200 meters between Basel and Strasbourg; 230 to 250 meters between Strasbourg and Mannheim. The river's length was reduced by 82 kilometers: 31 kilometers were removed in the stretch from Basel to the Lauter mouth (14 percent of the original 218.5 kilometers), the remaining 51 kilometers between the Lauter mouth and Mannheim (37 percent of the original 135.1 kilometers).[11] The new bed was designed to hold a maximum one thousand cubic meters of water per second, more than enough to contain the Rhine under normal conditions. Under high-water conditions (up to two thousand cubic meters per second), the Rhine was expected to overflow its artificial channel and pour into its old branches, loops, and oxbows. Whenever the river's water volume exceeded two thousand cubic meters, flood conditions would prevail. For the inhabitants of the Upper Rhine, this meant a significant reduction in the frequency of inundations, though not the elimination of the flood danger. Waterborne diseases all but disappeared, and a significant amount of land—around 10,000 hectares (or 100 square kilometers)[12]—was reclaimed for cultivation and for city growth.

Although the rectification did not begin until 1817, the Tulla Project originated in 1812 as a memorandum prepared on behalf of Baden for the Magistrat du Rhin, Napoleon's proto-Rhine Commission. The Magistrat approved the project, but by then the French Empire was beginning to falter (Napoleon was caught up in his ill-fated Russian campaign), and the Magistrat went defunct before work commenced. Tulla presented his plan anew in 1815, this time directly to Baden's riparian neighbors—the Bavarian Palatinate, Hesse, and Bourbon France—and in the course of the next twenty-five years four separate diplomatic treaties were signed. The first two, the Treaty for the Straightening of the Rhine from Neuburg to Dettenheim (1817) and the Treaty for the Rectification of the Rhine Flow Between the Neupfotz Cut and the Frankenthal Canal (1825), covered the entire length of the Baden-Palatinate border. The third, the Hesse-Palatinate Treaty, extended the rectification work downstream from Oppenheim to Worms; Tulla's colleague and collaborator, the Hessian engineer Claus Krönke, oversaw this part of the project. The fourth agreement, the Rhine Boundary Treaty (1840), signed by France and Baden, allowed the rectification work to continue upstream along the Alsatian border to Basel; it took over twenty years to negotiate, because neither side could resist sparring for control over every island and sandbar.[13]

From an engineering perspective, the Tulla Project can be divided into two phases. Phase I focused almost exclusively on the elimination of loops and oxbows along the meandering Baden-Palatinate border (see figure 3.1). The pick-and-shovel construction work was labor intensive: on average four thousand workers were needed for each cut. Typically the workers dug a ditch, 18 to 24 meters wide and about 6 kilometers in length, along a predetermined path that was to become the new riverbed. Once the ditch was dug, Rhine water was allowed to begin pouring through it. The velocity of the river itself was sufficient to widen and deepen the cut, and to achieve a stable new bed within a few years. Phase II focused on the removal of branches and islands, mostly along the braided Baden-Alsace and Hesse-Palatinate borders. Some 9 million cubic meters of earth and stone were used to construct new banks and diversion structures along the Basel to

Strasbourg stretch. About 3 billion cubic meters of earth, mostly islands and sandbars, were removed from the riverbed to ensure a uniform flow (2218 islands were eliminated between the Swiss and Hesse borders alone).[14]

From the outset, the Tulla Project was beset with difficulties. On several occasions, the Baden government had to send in military troops to oversee the construction, because some farmers and fishermen (most dramatically at Knielingen) opposed the removal of the river's oxbows and loops out of fear that valuable land and fishing holes would be lost. Opinions about Tulla varied greatly from village to village: locals tended to favor the cuts if their village would be farther from the river than before, but opposed them if their village would finish nearer the river. Fear of flooding evidently loomed large in their minds. Adding to the unrest, some of the initial cuts were undertaken using corvée labor (compulsory work based on the old feudal system), a practice quickly abandoned in light of the discontent it caused. Fears over the rectification work subsided once it became clear that the cuts reduced the flood danger all along the Upper Rhine while opening up new land for cultivation—all without negatively affecting (in the short run at least) the number of fish the river could support.[15]

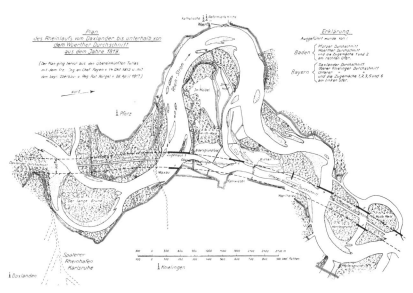

Figure 3.1 The Tulla Project on the Upper Rhine: blueprint from 1819 for the Daxlander and Woerther cuts (Source: Wasser- und Schiffahrtsdirektion Duisburg, *Der Rhein: Ausbau*, 102)

A less immediate but more enduring problem with the Tulla Project was channel-bed erosion. By shortening and straightening the Upper Rhine, Tulla deliberately steepened its gradient to accelerate the river's velocity. The swifter flow brought two advantages: the river dug itself a deeper and more stable bed with higher flood banks; and the water table dropped, turning "swamp" and "bog" into arable land. Unfortunately, the rectification work also brought with it several unintended consequences, which became more and more obvious as time passed. Tulla had assumed that the Rhine would cut a relatively uniform bed, deepening the river bottom by about a meter. But erosion patterns proved far more complex and unpredictable. Just below Basel, at the Isteiner Klotz (a cliff outcropping), the Rhine dug down 7 meters until it hit sheer rock. The rocks created rapids, making it increasingly difficult for ships to navigate upstream to Basel. Elsewhere along the French-Baden frontier (especially from Basel to Strasbourg), the bed sank far more than expected—nearly 6 meters at Rheinweiler and Neuenburg (see figure 3.2). The rectification work along the Baden-Palatinate border, moreover, did not produce as straight a river as planned: engineers sliced directly through the oxbows without completely eradicating the curve, so water sloshed back and forth from bank to bank as it flowed, creating new gravel reefs and sandbanks. Below Mannheim, instead of cutting a deeper bed, the Rhine was becoming plugged with debris from upstream—all to the detriment of Rhine transportation and commercial development.[16] Subsequent generations of engineers would therefore have to revisit the Upper Rhine to rectify what Tulla had supposedly ameliorated.

Tulla did not live to see his project completed or to address these unforeseen problems. He died in 1828 while undergoing medical treatment in Paris, and was buried there at Montmartre cemetery. A commemorative stone in Karlsruhe, his town of birth, reads: "Johann Gottfried Tulla: Dem Bändiger des wilden Rheins" ("Tamer of the Wild Rhine").[17]

Prussia on the Middle and Lower Rhine

Three decades after the Grand Duchy of Baden began "taming" the Upper Rhine, the Prussian government undertook the Prussian Navigation Project (1851–1900) on the Middle and Lower Rhine stretches between Bingen and the German-Dutch border. Prussia's two primary Rhine possessions, Rhineland and Westphalia, were artifacts of the French Revolution and the Congress of Vienna. A third possession, the Duchy of Nassau, was added after the Austro-Prussian war of 1866. Since Prussia controlled both sides of the Rhine from the borders of Alsace-Lorraine and Hesse in the south to the Dutch border in the north, the construction work was accomplished without the need for complicated international diplomacy. But the project did require intra-Prussian and intra-German cooperation, and was therefore part and parcel of Prussia's larger nation-building endeavors.

The Prussian Project got under way in 1851 in Koblenz with the creation of the Rhine River Engineering Administration, which brought hydraulic engineers from the Rhineland, Westphalia, and Nassau together under one roof. Its first director, Eduard Adolph Nobiling, had worked on the Elbe, Mosel, Saar, and Ruhr rivers before being entrusted with the task of enhancing the Rhine's navigability through massive reengineering and canalization. He was an adept administrator and served with great distinction for a quarter century until near-blindness forced his retirement in 1877. "I tried never to forget the interests of the humans affected by my decisions, recommendations, plans, and directives," he stated at the end

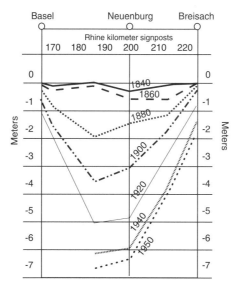

Figure 3.2 Bed erosion on the Upper Rhine between Basel and Breisach, 1828–1950. The Grand Canal d'Alsace was designed to halt the erosion while making this stretch navigable again. (Adapted from Tittizer and Krebs, eds., *Ökosystemforschung*, 30)

of his career. "Everything was designed to improve trade, industry, traffic, and agriculture as much as possible." A loyal Prussian monarchist, he was greatly embarrassed when his nephew, Karl Eduard Nobiling, attempted to assassinate Kaiser Wilhelm I in 1878 (an event that prompted Nobiling to change the family name to "Edeling" before his death).[18]

Nobiling's basic engineering challenge was to maximize navigational efficiency by maintaining a uniform flow at a predictable velocity. For this, he needed the right combination of width and depth for each stretch of the river. The blueprints set minimum depths of 2 meters from Bingen to St. Goar, 2.5 meters from St. Goar to Cologne, and 3 meters from Cologne to Holland. To achieve this depth, the river would require a width of 90 meters from Bingen to Oberwesel, 120 meters from Oberwesel to St. Goar, and 150 meters from St. Goar to Holland. The Lower Rhine (much like the Upper Rhine) had many meanders and braids, so Nobiling and his staff began with a series of Tulla-like cuts designed to shorten and steepen the stream. These cuts reduced the distance between Bonn and Pannerden (just north of the German-Dutch border) by 23 kilometers. At the same time, islands were removed or fastened to their adjacent banks, sandbars and other navigational hindrances were dredged, embankments were fortified, and harbors were added.

The Prussian Project, however, was not an imitation of the Tulla one. Whereas flood control had been Tulla's chief goal, ease of navigation was Nobiling's. Therefore, the Lower Rhine's bed was dredged extensively in order to create and maintain uniform shipping lanes. Nobiling's most original contribution was the design of a new type of wing dam (or groyne), an artificial barrier that extends into the river like a pier to shelter the banks from wave erosion and to direct the flow during low-water periods toward the middle of the stream so as to maintain a minimum shipping depth. These wing dams—concave structures averaging about a meter wide and 6.5 meters long—are still one of the most visible features on the Lower Rhine's banks today and the tiny ripples they produce are known as "Nobiling waves."[19]

The Middle Rhine, which flows through the narrow valleys and cliffs of the Rhenish Slate Mountains, created unique engineering problems for Nobiling and his staff (see figure 3.3). Rocks had to be blasted from the riverbed, sand removed from the shores, and canyon walls modified to reduce navigational dangers. There were eighty-five underwater rock barriers along the canyon walls and river bottom in the Bingen to St. Goar stretch alone,

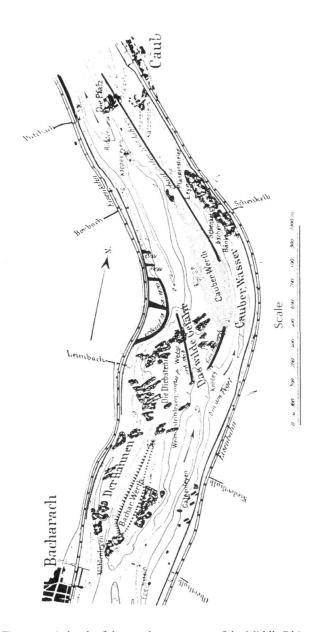

Figure 3.3 A sketch of the treacherous curves of the Middle Rhine circa 1860. Along the narrow passageway (near the middle of the picture) is written "Das wilde Gefahr," meaning "wildly dangerous." (Source: CHR, *Der Rhein unter der Einwirkung des Menschen*, 129)

including the treacherous curve at the Lorelei. Between 1851 and 1879, workers blasted out more than 30,000 cubic meters of rock, laboriously carving out a narrow shipping lane. When it proved inadequate, the process was repeated from 1890 to 1898 with better equipment. This time nearly 260,000 cubic meters of rock were removed.[20]

The Bingen reef posed a similar challenge. The reef created rapids when the water flowed at average depths, and acted as a natural dam during low-water periods. A 9-meter-wide gap (the Binger Loch) had been bored into the reef during the Middle Ages, but this hole was too narrow for modern boats and it still left the river too shallow to allow year-round shipping. From 1830 to 1832, a Prussian engineering team managed to drill a 23-meter-wide gap with the use of gunpowder and a large chisel-like device; but the construction of ever larger ships rendered this hole obsolete within a few decades. Under Nobiling's direction, a new gap was cut with the use of dynamite and dredgers in the early 1870s. Twenty years later the original gap was widened to 30 meters to allow two-way shipping at all times. After that, the reef no longer posed a serious hindrance to Rhine commerce, though modifications did have to be made in 1925–31 to accommodate the increasingly larger ships.[21]

Nobiling and his staff had to pay special attention to the tributaries of the Middle and Lower Rhine, all of which had a tendency to build up sandbanks and gravel pits at their mouths, impeding traffic on the Rhine. The Lahn, for instance, naturally entered the Rhine at a near ninety degree angle. It had to be widened and its mouth relocated downstream to reduce the angle of merger and thus the sand-and-gravel banks. The Sieg, which meandered wildly at its mouth near Bonn, had to be similarly rectified. The Mosel, at Koblenz, required the most extensive rectification work. Its mouth was naturally wide and shallow, rendering it unusable for shipping during low-water periods and prone to flooding under high-water conditions. An initial effort was made during 1860 and 1861 to fortify the left and right banks with stonework and wing dams. When that proved inadequate, more work was done in 1875–77, until the Mosel mouth began to resemble the Ehrenbreitenstein fortress located directly across from it on the Rhine's right bank.[22]

In 1893, the German government found a use for the newly fortified triangle (the "Deutsches Eck") at the Mosel-Rhine mouth—as a pedestal for a larger-than-life statue of Kaiser Wilhelm I and his horse. It was paid for by grateful Rhineland entrepreneurs to remind visitors of Prussia's Rhine imperium, and it became the object of much praise and derision. Kurt Tucholsky, passing by it many decades later, called it "a wonderfully Wilhelminian artifice of an artwork" ("ein herrliches, ein wilhelminisches, ein künstlerisches Kunstwerk").[23] He meant the monument but he might as well have meant the river, for the new Rhine was beginning to resemble the new Germany: fortified, centralized, uniform, and harnessed to the needs of its Prussian masters.

The Netherlands and the Rhine Delta

The Dutch did not stand finger-in-dike while others drilled and blasted upstream. Hydrology was for them politics by other means. With over half of the Netherlands lying below sea level, anything that affected the Rhine-Meuse delta affected their national security.

From an engineering standpoint, the Delta Rhine's drainage pattern had two natural defects. First, its two main channels—the Waal and Nederrijn-Lek—were too shallow and lacking in gradient to carry sediment and ice floes efficiently to the North Sea. Second, the various delta mouths were too narrow to keep all the water flowing in channels during peak discharge periods. The Tulla and Prussian projects exacerbated these natural deficiencies because they opened up the upstream stretches, thus allowing more sediment and debris to flow uninhibited into the delta and clog the channels. The partial confluence of the Waal (the main branch of the Rhine) and the Meuse rivers at Heerewaarden and again at Loevestein also posed a huge problem. When the Waal ran high, the Meuse tended to become so backed up that it would regularly overspill its banks. There were other problems on other parts of the delta drainage system as well. Flooding in 1809, 1820, 1824, 1855, and 1861 made it clear that the Dutch government would have to undertake the same kind of comprehensive flood-control and navigational improvements on the Delta Rhine that were under way upstream.[24]

Work would have commenced earlier, but a long period of political instability in the Netherlands delayed it repeatedly. The French Revolution, which swept away the *ancien régime* Netherlands just as it swept most other political entities on the Rhine, had ushered in a difficult period of Napoleonic rule followed by an equally difficult period of the United Netherlands (the Congress of Vienna's ill-fated Dutch-Belgium union). Napoleonic France established the Comité Central du Waterstaat in 1809, similar in scope and purpose to other French-inspired river commissions. Though it lasted only a few years, it stimulated cooperation among the Dutch provinces, and in subsequent decades a variety of special commissions hammered out a comprehensive delta flood-control plan. Work finally began in earnest in 1850, after Belgium had secured its independence (1830–39), and the 1848 revolutions had fizzled.[25]

A new central agency, the Rijkswaterstaat (Department of Public Works and Water Management), oversaw almost all of the work on the Delta Rhine. Its first major undertaking, the Merwede Project (1850–1916), concentrated on a section of the commingled Waal-Meuse waters between Gorinchem and Dordrecht. Workers began by constructing a new riverbed—the Nieuwe Merwede (or New Merwede)—for the middle portion of the Merwede river, which in its natural state was more a series of interflowing creeks than a river (see figure 3.4). The new bed provided the Waal and Meuse with a deeper and wider conduit to the sea, one that worked well for flood control as well as navigation. The stretch immediately downstream from the Nieuwe Merwede—known as the Boven Merwede (Lower Merwede)—was more problematic. Its bed had grown so wide (600 meters) from transporting the combined Waal and Meuse waters that during low-water periods it easily became choked with sandbars and therefore largely unnavigable. To rectify this problem, the Rijkswaterstaat dug a new mouth system for the Meuse—known as the Bergse Maas—which greatly reduced the amount of commingling between the Waal and Meuse, and opened up traffic to the sea via the Hollandsch Diep. Once Meuse waters no longer encumbered the Boven Merwede, engineers were then able to reduce its width so that it became navigable year-round.[26]

As the Ruhr region grew into an important industrial and commercial center in the second half of the nineteenth century, the Rhine Commission put more and more pressure on the Netherlands to improve its waterways and sea routes to handle the greater traffic and increasingly larger ships. The Dutch responded with a variety of initiatives. The first, the Nieuwe Waterweg Project (1860–72), required the construction a 35-kilometer "New Waterway" to link the Waal to the North Sea via Rotterdam. Following a plan developed by Pieter Caland, engineers cut a trench through the Hoek van Holland, dammed the old river channel, and let the river current forge a new Waal mouth. Dredgers were then used to deepen the river channels, creating the preconditions for Rotterdam's phenomenal rise as the world's largest harbor. More Rhine-related projects followed. From 1875 to 1906, engineers redesigned the Waal and Bovenrijn (the short stretch of the Lower Rhine between the German border and Pannerden canal) so that they would maintain a minimum depth of 3 meters. These alterations made it easy for large ships and barges to travel freely back and forth between Ruhr and Rotterdam. At the same time the Nederrijn-Lek system was reengineered to maintain an alternate shipping route at a minimum depth of 2 meters. Then, from 1920 to 1932, engineers dammed the Zuiderzee, transforming the IJsselmeer (a minor Rhine mouth) into an estuary. The widening and canalization of the Meuse (1918–42) and the

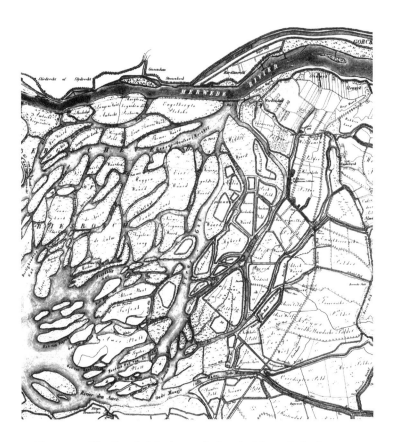

Figure 3.4 The Dutch delta in 1821 before the Merwede Project. Much of the Delta Rhine consisted of a main channel (shown here at the top) and a multitude of interlocking creeks. (Source: Wiebeking, *Atlas*, Section X)

Nederrijn (1954–69) followed. Meanwhile, in the 1930s the Amsterdam-Rhine canal (1931–38) was built, providing Rhine ships with yet another route to the North Sea.[27] Finally, after a coastal storm devastated the Netherlands in 1953, the Dutch undertook their greatest engineering feat of all—the Delta Plan—designed to shield the country's coastline, harbors, and rivers as much as feasible from the North Sea. While not a specifically Rhine project, it did necessitate the construction of several dams (most notably the Haringvlietdam) on the various Rhine mouths, with negative consequences for migratory fish.[28]

The Alpine Stretches

While others worked downstream, Swiss engineers began rectifying the Alpenrhein, Aare, and High Rhine in cooperation with the neighboring states of Austria, Liechtenstein (a minor player), and Germany. Navigation was not a major concern. The main goals were flood control and hydroelectricity production.

The Alpenrhein tributary system (upstream from Lake Constance) was by nature unpredictable. Its flow was irregular and its deposition pattern haphazard, rendering human settlement in the Swiss canton of Grisons and the Austrian province of Voralberg exceedingly precarious. Mining and textiles were the prime industries, agriculture and livestock less so due to frequent inundations. The timber industry boomed in the eighteenth century, giving rise to more and more lumberjack communities situated directly on the floodplain. Meanwhile, deforestation from lumbering accelerated river erosion, thereby increasing the severity of floods. The floods of 1868 and 1871 brought the three riparian states—Switzerland, Austria, and Liechtenstein—to the negotiating table in search of a mutually acceptable rectification plan. The Swiss were the driving force behind these efforts, since they were eager to develop the Alpenrhein's economic potential. The Austrian government still tended to view the Alpenrhein (which lay far removed from its Danube commercial centers) as a fringe region. It therefore took more than twenty years—and several more floods (1888, 1890, 1892)—before the Swiss-Austrian State Treaty of 1892 was finally negotiated and ratified.[29]

Since none of the Alpine states belonged to the Rhine Commission at the time (Switzerland would not join until 1919), they created their own International Rhine Regulation Commission to oversee the Alpenrhein Project (1895–1973). The initial plan was to provide a riverbed 110 meters wide, which it was hoped would steepen the river's slope sufficiently to give its waters enough momentum to transport its sand and gravel into Lake Constance. The project was carried out in two phases. First came the Fußach cut, which reduced the Alpenrhein's length by 7 kilometers, steepened its slope, and moved its mouth 8 kilometers westward. Many years later came the Diepoldsau cut, which removed a large curve and shortened the Alpenrhein by another 3 kilometers. Engineers also regulated the river above Fußach with dams and embankments, and harnessed the stretch between the Ill river (not to be confused with the identically named river in Alsace) and Diepoldsau.[30]

Unfortunately, the Alpenrhein had not been steepened sufficiently and so its bed continued to silt up with sand and debris. By the 1930s, the riverbed had risen nearly 2 meters at Diepoldsau, which meant that the river's water was now flowing well above the surrounding landscape, presenting a new flood danger to the nearby communities. In 1954, after a new Swiss-Austrian treaty was signed, engineers reworked the Diepoldsau and Fußach cuts, and built narrower flood dams all along its banks. They also narrowed the river to 70 meters, thereby increasing the velocity enough for the water to shove its glacial debris into Lake Constance. This project worked much better and today Lake Constance swallows 3 million cubic meters of sand and gravel, and 50 thousand cubic meters of glacial rubble, each year—enough to the keep the Alpenrhein from silting up.[31]

Regulation of the Alpenrhein and the Lake Constance delta constituted the main, but not the sole, task of the Alpine engineers. As part of their flood-control efforts, they dammed and reinforced nearly all the rivulets and streams of the Vorderrhein, Hinterrhein, and Alpenrhein. These measures helped open up land to cultivation and stock raising, and provided new employment opportunities for the lumber and mining communities, many of which had experienced hard times as the nearby forests and mines became depleted. Even more important than reclaiming land was the creation of a new industry: hydroelectric power (or "energy gold" as the miners called it). The Vorderrhein's steep northern valleys (near the Aar massif) proved ideal for energy production, as did much of the Hinterrhein. The first plants were built at Ragaz (1892), Frauenkirch (1894), Glaris (1899), and Flims (1904), mostly for local consumption. Then, in 1910, Zurich built a plant at Sils, along with the long-distance power lines needed to "export" the energy 150 kilometers to the west. With a yearly production of 130 million kilowatt hours, it was the largest of its kind at the time. Other projects followed in quick succession, and by 1990 the Swiss-controlled section of the Alpenrhein basin was bedecked with 28 reservoirs and 61 hydro-electric plants, which collectively accounted for 15 percent of Switzerland's yearly hydroelectric production.[32]

While work progressed on the Alpenrhein, the Swiss also undertook the Jura Correction Project (1867-1973) on the Aare, the Rhine's largest Alpine tributary. Since the Aare system lay wholly within Switzerland, there was no need for international negotiation, though as was often the case the team of engineers and consultants was international. Tulla had advocated rectifica-tion work on the Aare at the beginning of the nineteenth century, but it was not until many decades later that Richard La Nicca drew up detailed blueprints. As usual, it took a flood—in 1865—to prod the Swiss govern-ment into action. The basic plan was to turn the natural mountain lakes into reservoirs. In 1878, engineers redirected the Aare river so that it began to flow into Lake Biel via the Aarberg-Hagneck canal. The old arm of the Aare from Aarberg to Büren was thereby cut off and Lake Biel utilized as a natural reservoir. The Broyce river, linking Lake Morat to Lake Neuchâtel, became the Broyce canal, while the Zihl river, linking Lake Neuchâtel to Lake Biel, became the Zihl canal. The Zihl river below Lake Biel became the Nidau-Büren canal; it took waters from Lake Biel back to the original Aare riverbed at Büren. A follow-up plan expanded the carrying capacity of the canals and widened the Aare between Solothurn and the Emme mouth. As elsewhere, these changes were designed primarily for flood control but they also produced hydroelectric power and helped open up former wet-lands for cultivation.[33]

Without doubt, the most significant Alpine engineering work was done on the High Rhine stretch between Lake Constance and Basel along the Swiss-German border. The High Rhine's rock walls and steep slope (150 meters over a 142-kilometer stretch) made it an ideal location for the production of hydroelectricity, and all of the affected riparian states and cantons—Baden, Aargau, and Basel—worked in concert to develop its potential. The first hydroelectric plant, a small one at the Swiss town of Schaffhausen, was installed in 1866. Then in the 1890s the era of large hydrodams began in earnest: Rheinfelden (1898), Augst-Wyhlen (1912), Laufenburg (1914), Eglisau (1919), Ryburg-Schwörstadt (1931), Albbruck-Dogern (1934), Reckingen (1941), Rheinau (1956), Säckingen (1966), Birs-felden (1954), and Schaffhausen (1963).[34]

The master plan for hydroelectric development on the High Rhine was known as Project Free Rhine—without a hint of irony.[35]

The Upper Rhine Revisited

Engineering work on the High, Middle, and Lower Rhine forced the German and French governments to redo the Upper Rhine. The Tulla Project had concentrated on flood control, not navigation. In fact, the rapids that had inadvertently resulted from the original rectification work, especially at Istein, had all but eliminated ship transport between Mannheim and Basel. Flood control, in other words, worsened the Upper Rhine as a transportation route, a situation that the Rhine Commission would not tolerate forever. Initially, the railroads stepped in to fill the void—Baden developed one of Germany's most extensive train networks—but rail was far more expensive than barge for the transport of *bulk* goods. Therefore, the Swiss government put increasing pressure on Germany to redo the Upper Rhine so that ships and barges could reach Basel. In 1901, the Agreement between Baden, Bavaria, and Alsace-Lorraine of the Regulation of the Rhine from Sondernheim to Strasbourg was signed. A follow-up treaty in 1924 extended this work to the Mannheim-Strasbourg stretch, and a later third treaty to Basel.[36]

The Honsell-Rehbock Project (1906–36)—so named in honor of the two Baden engineers, Max Honsell and Theodor Rehbock, who oversaw the work—was in many ways a reengineering of the Upper Rhine to bring it into conformity with the Prussian and Dutch projects downstream. The goal was to create a riverbed 2 meters deep and 88 meters wide that would permit all-year navigation from Mannheim to Strasbourg and eventually to Basel. Dikes, dams, and wing dams were needed, as were reinforced riverbanks and harbors. To assist in the rectification work, Rehbock even built a 60-meter model of the river, the "Rheinhalle," in his Karlsruhe laboratory, which he utilized to test and predict sediment buildup. Special attention had to be paid to the difficult problem of maintaining the water level, because the Upper Rhine was supplied mostly by Alpine runoff and was therefore prone to seasonal droughts. As the project's first phase neared completion in the 1920s, it became possible for large ships to travel beyond Mannheim (the endpoint of Rhine shipping during most of the nineteenth century) to the new ports of Karlsruhe and Kehl/Strasbourg for most of the year.[37]

However, the route farther upstream along the Alsatian-Baden border (from Strasbourg to Basel) remained fraught with innumerable hazards, natural and political. Baden and Switzerland had agreed in principle as early as 1879 to make the Neuhausen-to-Basel stretch navigable, but implementation was slow in coming.[38] One obstacle was the Istein rapids, which would require an elaborate lock system to circumvent. A related problem was the water table in Baden. Any rectification work that entailed extensive canalization of the riverbed would inevitably lower the water table and potentially endanger Baden's agricultural productivity. A final obstacle was the city of Mannheim. Ever vigilant against competition from Basel's harbor, city leaders used their influence in the Rhine Commission (which at that time was almost wholly dominated by German states) to slow down the rectification work upstream. Freight and passenger service did finally begin on the Strasbourg-to-Basel stretch in 1905, but the steam companies could not guarantee year-round shipping and therefore could not compete with rail service. In all of 1913, for instance, only 97,000 metric tons went on the steamers, a minuscule fraction of the total German-Swiss trade.[39]

After World War I, however, the Germans no longer had the diplomatic leverage to hinder the rectification work between Strasbourg and Basel: Articles 354-362 of the 1919 Versailles Treaty gave France direct control over the left bank (Alsace) as well as the legal right to develop the Upper Rhine's hydroelectric potential. Out of Versailles came the Grand Canal d'Alsace (1921–59), designed by René Koechlin. He envisaged the construction of eight hydroelectric plants spread roughly evenly apart from Basel to Strasbourg. The basic idea was to divert nearly all of the river water from its original bed, and channel it into a canal-and-lock system running parallel to the river's left bank. Aside from producing hydroelectric power, the Grand Canal was to serve as a transportation link between Basel, Mulhouse, Colmar, and Strasbourg. It would also promote greater commercial ties between the Rhine and Rhône river systems, and spawn industrial growth—French industrial growth—in the newly reconquered Alsace-Lorraine region.[40]

The dilemma for the French was that German engineers were able to make a compelling case against the Grand Canal on purely technical grounds. The idea of constructing a "Rhine Lateral Canal" (as the Germans call the Grand Canal) was not a new one: Prussia and Baden had discussed it twenty years earlier as a potential spur to navigation and industry on the Upper Rhine. Plans had been drawn by Koechlin, costs estimated, the pros and cons weighed again and again. The project had never been realized, however, largely out of concern that it would endanger the water table in Baden. The Tulla Project had already disrupted the natural hydrology of the Upper Rhine and substantially lowered the groundwater level. As a consequence, Baden farmers were already highly dependent for their livelihood on irrigation water, pumped either directly out of the ground or from the Rhine. If Rhine waters were diverted into a lateral channel, farmers could no longer nurture their crops and the land would remain fallow. There were even some exaggerated fears—grist for the Nazi propaganda mill—that canalization of the Upper Rhine would set in motion the "steppification" of Germany.[41]

As of 1932, France had built only one hydroelectric plant, at Kembs. Germany and Switzerland supported this project, as did the Rhine Commission, because it included locks for bypassing the Istein rapids. Plans to build more dams were scrapped in 1940 when Hitler brought all the riparian states except Switzerland and Liechtenstein under Nazi occupation and turned the Rhine into a Nazi river. Then, immediately after World War II, the French pulled out their blueprints anew and built three more hydroelectric plants and locks in rapid succession—at Ottmarsheim (1952), Fessenheim (1956), and Vogelgrün (1959)—in part with funds from the Marshall Plan. Franco-German rapprochement, however, put a halt to the final completion of the Grand Canal Project in its original form, especially once it became clear in the 1960s that the canal did in fact endanger water supplies, urban sanitation, and agricultural productivity in Baden (part of Baden-Württemberg after 1952). A new agreement, the Treaty for the Improvement of the Upper Rhine from Basel to Strasbourg (1956), left the already completed portion of the Grand Canal intact. However, the stretch from Breisach to Strasbourg/Kehl (site of the next four planned hydroelectric plants) was reengineered following the so-called Loop Solution, a plan worked out jointly by German and French engineers. The Loop Diversions allowed the Rhine to continue flowing in its original bed for most of this stretch. At four points, the water was diverted to hydroelectric plants and locks: Marckolsheim (1967), Rhinau (1963), Gerstheim (1967), and Strasbourg (1970). France agreed to this more-expensive solution only after the Germans agreed to the full canalization of the Mosel river.

The Loop Diversions helped to alleviate the water table problems in Germany, but not the erosion problems that had been confounding engineers since rectification work began in 1817. Hydrodams alter deposition patterns in two contradictory ways: the in-river reservoirs above the dams collect gravel, sand, and silt that would otherwise continue to flow downstream, thus diminishing the natural sediment load of the river; meanwhile, as it returns to the river, the water that pours out of the turbines below the dams scours the bed, thus greatly augmenting channel erosion at that point. In 1969, the French and German governments built two more in-river systems—a hydrodam at Gambsheim (1974) and a hydro-lock at Iffezheim (1978)—which were specifically designed to ease some of the erosion problems below the Loop Diversions. Then, in 1982, they began dumping millions of tons of gravel and rock into the river downstream from Iffezheim in order to compensate for the sediment loss caused by the upstream dams. So far it seems to have stabilized the bed, resolving a problem that had persisted since the days of Tulla.[42]

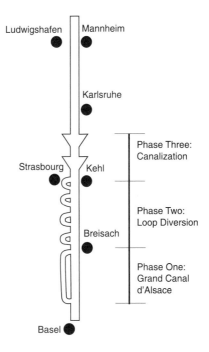

Figure 3.5 Schematic diagram of the Upper Rhine today. Between 1921 and 1978, engineers experimented with a variety of techniques designed to balance the needs of hydroelectric development, navigation, agriculture, and erosion control. (Source: CHR, *Der Rhein unter Einwirkung des Menschen*, 83)

Flood Control Revisited

Flood control proved to be even more intractable than erosion control on the new Rhine. By and large, engineers successfully eliminated the problem of severe flooding along the Alpine, High, and Upper Rhine stretches (roughly between Chur and Bingen), where the river's natural fickleness once made human settlement precarious. On the Alpenrhein, hundreds of small dams now regulate the meltwater streams; on the Aare system (including the Reuß and Limmat), nine natural lakes are now harnessed to serve as river reservoirs; and on the High Rhine and Upper Rhine, a total of twenty-one dams help even out the river's day-to-day flow. However, some of these projects have increased the flood danger on the Middle, Lower, and Delta Rhine—as the four "hundred-year" floods of 1983, 1988, 1993, and 1994 recently demonstrated beyond a shadow of doubt. The reason for this incongruity is that the flood-control measures undertaken to regulate the river between Chur and Bingen were not built in harmony with the navigation-related measures undertaken to regulate the river between Basel and Rotterdam. Not surprisingly, most of the problem is centered on the area of overlap—the Upper Rhine between Basel and Bingen—where the in-river constructions designed to reduce flooding and improve navigation (the Grand Canal d'Alsace and Loop Diversions) have inadvertently embedded a faulty drainage regime into the river. Without a floodplain at the base of the Alps to absorb some of the flow, Alpine water has nowhere else to go but downstream en masse. Put simply: more upstream water flowing more quickly translates into more downstream flooding.

A handful of hydraulic engineers—notably Freiherr van der Wijck and Fritz André—recognized the danger from the outset. Their critiques, lost at first amid the fanfare of the Tulla Project, surfaced in full force in 1824 when a devastating flood struck the Middle and Lower Rhine. Van der Wijck's broadside, "The Middle Rhine and Mannheim in Light of Hydrotechnology," was a relatively mild-mannered critique of Tulla. André's broadside, "Remarks on the Rectification of the Upper Rhine and a Description of the Dire Consequences It Will Have for the Inhabitants of the Middle and Lower Rhine," was as provocative as it was alarmist. He charged that Tulla had willfully neglected well-known facts about the dynamics of watershed drainage in Central Europe. "Thaw winds," he wrote, "generally sweep through Germany and its neighboring states at winter's end, causing a general snowmelt and a swelling of the streams." The small Kinzig river, for instance, reached its maximum swell nearly three days before its larger neighbor, the Main river. The Main, in turn, peaked a full day and a half before the Rhine swell reached the Main mouth at Mainz. Similarly, the Neckar dumped its load into the Upper Rhine at Mannheim several days before the trunk river itself began to peak. "If the Upper Rhine is rectified as completely as envisaged by the Tulla Project," he predicted, "it will create a shorter and swifter current that will cause all of Germany's rivers to reach their peak discharges at or about the same time. The result will be a significant increase in flooding." [43]

The critiques of van der Wijck and André (and similar ones by other water experts) alarmed the downstream governments enough for them to take diplomatic action. Prussia began by petitioning the German Confederation (the intra-German forum that replaced the Holy Roman Empire) to use its powers to suspend the Tulla Project pending a scientific inquiry. The Dutch government became a party to the suit in 1829, while the state of Hesse joined in 1832. Ratification of the Mainz Acts (the first revision of the Vienna accords) in 1831 lent political and moral gravity to their petition: Article 18 explicitly forbade any rectification work that did not have the unanimous support of all the riparian states (though it remained legally unclear whether the acts could be retroactively applied to projects already under way). Under these circumstances, Baden and Bavaria found it prudent to suspend the rectification work until it could be determined whether God or Tulla was to blame for the 1824 flood. The German Confederation then turned the matter over to a special commission composed of representatives from all the affected German parties (with Prussia representing itself and the Netherlands). It set up headquarters in Speyer in early 1831. [44]

In pleading their case, the Prussian and Hessian diplomats and engineers focused on the long-term implications of river manipulation, arguing that intergovernmental cooperation was the sine qua non of rectification work. "Every river needs to be viewed as a single entity from its source to its mouth," stated a Prussian legal brief, in language that has a remarkably modern ring to it. "It belongs to all of the states through whose territories it naturally flows. Every artificially induced change in its natural conditions that has an impact across the state borders will affect the property of the adjoining state. Negative impacts entail destruction of property and therefore violate interstate law." [45] In defending themselves, Baden and Bavaria's diplomats and engineers (Tulla among them, until his death in 1828) never contested the right of downstream states to halt projects that caused damage to their land and property. Instead, they simply disputed the notion that the handful of cuts that had been made since 1817 could possibly have caused the 1824 flood. They pointed out, quite correctly, that 1824 had been an especially wet year that would have caused flooding whether or not the cuts had been made. They also disputed the long-term implications of the rectification work: Tulla's blueprints called for carefully designed riverbanks that would continue to allow water to inundate the Rift valley during flood conditions and thus not become funneled downstream.[46]

The German Confederation's special commission met ten times over a six-week period in 1832 to discuss and evaluate the legal briefs and engineering arguments. But after numerous discussions and expert testimony from representatives on both sides of the case, the commission was forced to concede that it lacked sufficient information to render a verdict. Most troublesome was the lack of consensus among engineers as to the likely consequences of straightening the river. "Theories were pitted against theories, calculations against calculations," Honsell noted in his historical account of the dispute, "and the commission ended the negotiations without reaching agreement on a single aspect of the dispute."[47]

Stalemate on the legal front did not automatically have to translate into a full-scale retreat by the Prussia-led forces. Had the downstream states continued to mount a concerted diplomatic initiative against the Tulla plan, they might well have succeeded in influencing subsequent events in a manner that would have provided them with greater protection from upstream flooding. Periodic monitoring would certainly have helped to fill in the missing data so that subsequent generations of engineers and jurists could have assessed the Rhine's changing hydraulic conditions in a more clear-cut manner. But instead of turning up the pressure, Prussia, Hesse, and the Netherlands tacitly accepted the commission's final report as a defeat and thus stood silently by as the Tulla Project resumed in 1832. All they secured by way of compromise was an agreement from Baden and Bavaria to reduce the total number of cuts by four—and even that turned out to be a temporary concession.

There were several reasons why the downstream governments decided not to pursue the flood issue after 1832. To begin with, the anti-Tulla coalition was never as strong as it appeared on paper. Dutch authorities had little to say in the discussions, save placing their stamp of approval on Prussia's initiative. The state of Hesse, meanwhile, had joined the case at the last moment and then only half-heartedly. Claus Krönke, Hesse's chief water engineer, was a close friend and colleague of Tulla's. Even in the midst of the debate, he was preparing to oversee a new phase of the Tulla Project along the Hesse-Palatinate border between Oppenheim and Worms. The Prussian government too was acting in a disingenuous manner: as its diplomats argued the legal case against Baden and Bavaria, its engineers were preparing to drill a 23-meter gap in the Bingen reef to ease shipping on the Middle Rhine. The Rhine Commission entered the fray as well: "The general feeling is that the cuts should continue," the Prussian representative bluntly told his government in 1832, after returning from a meeting with his fellow commissioners.[48]

The flood issue resurfaced briefly in the 1880s, this time in the wake of back-to-back floods in November and December 1882. But by then rectification work was well under way everywhere on the river and flooding was not the hot-button diplomatic issue of yesteryear. To be sure, German officials were sufficiently concerned to ask Honsell (in his capacity as head of the German Central Bureau for Metereology and Hydrography) to issue a report on the causes of the 1882 floods. But his reassuring analysis put their concerns to rest. Honsell laid the blame on a series of unusual weather patterns that had caused the Rhine's four largest tributaries—the Mosel, Main, Aare, and Neckar—to overburden the trunk system. The November flood had occurred because the Main and Mosel swells had reached Koblenz simultaneously at a moment when the Middle and Lower Rhine was already running high. The December flood had occurred because the Aare's runoff reached downstream before the Neckar, Main, and Mosel waters had ebbed.[49]

Had they still been alive in 1882, van der Wijck and André would have been perturbed to find that the once-mercurial Upper Rhine between Basel and Worms had not flooded at all, even under these extreme high-water conditions. They would have pointed out that the water—which under natural conditions would have poured into the Rift valley—had instead been hurled en masse downstream where it met up with the tributary swells. Honsell, however, was not haunted by such matters; in fact, he proudly declared that the Tulla Project had completely protected the Upper Rhine from flooding!

Two centuries of nonstop hydraulic tinkering have given the Rhine a sleek and slender profile. Modified like no other large river in the world, it has become the axis of European water transportation. In 1810, a scant 328 ships traveled between the German states and the Netherlands, then as now the most commercially lucrative stretch of the river.[50] By the 1990s, the number of annual German-Dutch border crossings had surpassed 200,000 and the Rhine fleet stood at well over 13,000 vessels.[51] In 1840, downstream traffic still outpaced upstream traffic by a three-to-one ratio: 374,000 metric tons of goods (coal, wood, stone, wheat) passed the Dutch-German border heading downstream whereas only 128,000 tons of goods (coffee, cotton, seed, whale oil) passed the border heading upstream. By 1907, however, upstream traffic outpaced downstream traffic by a two-to-one ratio, with 15.8 million metric tons (ores, coal, wheat, wood) heading upriver and 7.3 million tons (coal, gravel, iron, stone, cement) down.[52] Total tonnage on the Rhine also increased year to year, rising from 10 million in 1825 to a staggering 288 million in 1978.[53]

Improvements in ship construction also made it increasingly cheaper and faster to transport goods and people, especially after engineers widened and deepened the river to accommodate vessels of the 3000-metric-ton class. Sailboats, whose number stood at nearly two thousand in 1880,[54] had all but disappeared from the river by 1935. So had the Holland rafts (commercial timbers bound together and floated downstream by skilled raftsmen). During the same period, the number of steam-powered ships and tugs jumped from two hundred to two thousand. In the 1920s, steamers began to give way to the new motorized freight ships, chief among them the diesels, which were powerful enough to transport bulk goods upstream on the new Grand Canal d'Alsace. Because of their versatility, these self-propelled vehicles have remained the most frequently used ships on the river to this day.[55] Tug-and-barge systems, first introduced in the mid-nineteenth century, also became an instant success, especially for hauling ores along the Duisburg-to-Rotterdam stretch. In the 1950s, they in turn gave way to the modern push-tow systems (a tug and six lighters), capable of transporting 16,000 metric tons of ore in one haul.[56]

These impressive statistics demonstrate how well the Rhine Commission accomplished its goal of turning the Rhine into an autobahn of global commerce and trade. But there was a price to pay for treating a river as just the sum of its parts. Each project tended to be tailored to the particular needs of a region, state, or locale at a particular time, with too little attention paid to the possible long-term consequences upstream or down. The Tulla Project unleashed massive erosion problems on the Upper Rhine, causing the Middle Rhine to clog with sand, gravel, and rock. The Prussian Navigation Project resolved that problem by opening up the Middle and Lower stretches so that the sediment could continue to the delta. This, in turn, forced the Dutch to reengineer their branches and mouths so that the debris could be funneled efficiently into the North Sea. Similarly, as ship traffic on the Lower and Middle stretches increased each year, the governments of Alsace, Baden, and Switzerland came under increasing pressure to undo the Tulla Project and reengineer the Upper Rhine so that ships could reach as far upstream as Strasbourg and Basel. This, in turn, exacerbated flooding on the Middle and Lower Rhine. Techno-fixes begat techno-fixes in the never-ending quest for the perfect canal-like river—even if they meant (as they did in many river stretches) a commitment to year-round dredging to keep the erosion problems in check, and even if they meant partial flood protection for upstream cities at the expense of those downstream.

In retrospect, it is easy to see that the riparian governments missed two major opportunities—in 1824 and again in 1882—to address an issue that would later come back at them with a vengeance: flooding. For in the long run André and his fellow critics understood the ramifications of Rhine engineering better than those who composed the blueprints and did the work. The first generation of engineers (Tulla and his co-workers) took the initial step when they removed the Upper Rhine's oxbows and curves, thereby shortening this river stretch and greatly accelerating its velocity. This altered the trunk river's runoff pattern in relation to its tributaries. The next generation of engineers (those who worked on the Honsell-Rehbock Project and later the Grand Canal d'Alsace) took the next step when they narrowed and canalized the channel and raised its embankments. More concerned with navigation than flood control, they completely severed the river from its floodplain in the interest of maintaining an adequate minimum flow. This eliminated the sponge-like function of the Upper Rhine's Rift valley. A final step was taken when the Rift valley began to fill up with farms, factories, and cities, not least to take advantage of the fact that the Upper Rhine had become fully navigable year-round for the first time in history. This made the transformations—for all intents and purposes—irreversible.

In essence, successive generations of Rhine engineers inadvertently built a ticking time bomb into the Rhine. In their haste to transport water quickly and efficiently downstream in one bed (following the principles of Renaissance engineering), and in their desire to construct the perfect canal (following the principles laid down by the Rhine Commission), they failed to take into account one of the most simple and well-known dynamics of river drainage. Moreover, both the German Confederation and the lower riparian states decided to drop the flood issue before the risks were fully known—so there was no scientific agency collecting and assessing flood data, no independent bureaucracy evaluating the engineering projects, and no judicial mechanism for keeping matters from spinning out of control.

Not all of the blame, however, lies with the Rhine Commission and its army of engineers. Each time they rectified a stretch of the river for transportation purposes, they also opened up new riverfront real estate, which in turn opened up new avenues for industrial growth. That spurred the growth of towns and cities, which created more pressure on engineers to turn over more of the river's floodplain to human development. The list of people involved in this free-for-all was a veritable who's who of European commerce and trade: shippers, farmers, miners, industrialists, loggers, warehousers, governors, mayors, urban planners as well as other municipal, state, and national officials. Towering above them all, however, was the one group that had the most to gain from the scramble for river land, river water, and river harbors: the coal industry of Rhineland and Westphalia.

In the German forest—
that forest which inspired the Erlkönig—
one now hears only a dismal fugue out of the
timeless reaches of the carboniferous.

— Aldo Leopold (1935)

Field and forest and stream and ocean
are the environment of life:
the mine is the environment alone
of ores, minerals, metals.

— Lewis Mumford (1934)

CHAPTER 4

The Carboniferous Rhine

$\mathcal{F}$ar more than textiles and locomotives—Lewis Mumford noted in *Technics and Civilization*—mining and metallurgy set the pace for modern industrial development. The use of coal as a fuel source for machine power, and the application of the steam engine to iron smelting, created the foundation for a "paleotechnic" civilization (a phrase Mumford borrowed from Patrick Geddes)—a civilization based on the wholesale extraction and consumption of age-old irreplacable resources. "The large-scale opening up of coal seams," Mumford wrote, "meant that industry was beginning to live for the first time on an accumulation of potential energy, derived from the ferns of the carboniferous period, instead of upon current income." Unlike wood, windmills, and waterwheels, coal was a nonrenewable resource. "The mine is the worst possible local base for a permanent civilization: for when the seams are exhausted, the individual mine must be closed down, leaving behind its debris and its deserted sheds and houses. The by-products are a befouled and disorderly environment; the end product is an exhausted one."[1]

Nineteenth-century mining caused a number of ecological disturbances to the surrounding air, water, and soil. Most immediately visible were the slag heaps piled up around the early mining sites and the smoke-filled skies that gave industrial regions their characteristic bleakness. Coal ores typically contain a multitude of impurities, notably sulfur. When burned in steam engines and boilers, and vented through smokestacks, sulfur returns to the earth as sulfur dioxide, the main ingredient of "acid rain." Half-dead forests and poor crop yields were common in regions where smokestacks dominated the horizon. So were lung diseases, which affected not only those who worked the coal shafts ("black lung") but also those forced to breathe the foul air around the factories. "The earth seems to have been turned inside out," wrote James Nasmyth, after visiting a British coal district in 1830. "Its entrails are strewn about, nearly the entire surface of the ground is covered with cinder-heaps and mounds of scoriae [slag]. The coal, which has been drawn from below ground, is blazing on the surface. The district is crowded with iron furnaces, puddling furnaces, and coal-pit engine furnaces. By day and by night the country is glowing with fire, and the smoke of the ironworks hovers over it."[2]

Mining also had a profound and lasting impact on water runoff and water quality. Coal and iron deposits are embedded in the thin layer of the Earth's crust (the lithosphere) that acts as a bedrock for underground streams, aquifers, and groundwater. When miners dig shafts and pits, they punch holes in this layer, permanently altering the manner in which a basin collects and sheds water. Moreover, because the shafts and pits fill up with surface and underground water, they have to be pumped continually to keep them serviceable; the deeper the shafts and pits, the greater the impact on the region's groundwater. It was common practice among mining firms (until recent environmental regulations were put in place) to dump this "mine water" into whatever lake or stream happened to be nearby—along with the coal dust, chlorides, phenol, sulfur (in the form of weak sulfuric acid), and other impurities. Further, as mines become depleted, they begin to collapse upon themselves, causing land subsidences on the surfaces above them. These depressions typically fill up with stagnant water, which can result in an upsurge in waterborne diseases such as typhoid and cholera. The needs of transport also often necessitated a resculpting of the hydrologic terrain: though railroads handled much of the traffic in raw and finished products from the mid-1800s onward, rivers and canals were the preferred modes for transport of bulk goods because the cost per ton was much cheaper.

Some of these environmental transformations were temporary, confined to certain regions at certain times during certain stages of their industrial development. Smoke-filled skies, for instance, began to disappear once scrubbers were added to smokestacks and especially once oil replaced coal as the universal fuel in the mid-twentieth century. Industrially advanced nations have also managed to eliminate the waterborne diseases that once plagued mining sites. But many of the transformations tended to create permanent blights on the landscape: altered stream channels, abandoned canals, slag-heap hills, cratered valleys, polluted soils, stunted forests, and abandoned farmland, among them.

The coal age brought with it industrial concentration in certain regions of Europe. It also fostered monopolistic practices and reinforced patterns of militarization in European economies and societies. Although wood, wind, and water power were widely available in Europe, the coal seams were confined to a handful of places, such as north-central England, south Wales, Upper Silesia (now part of Poland), Ukraine's Donets basin, and (most important for the Rhine) northwestern Europe from Lille in France to Liège in Belgium to Dortmund in Germany. Coal was difficult to extract from the earth, steam engines expensive to purchase, iron smelters costly to maintain. These economic constraints prompted large-scale solutions: huge steam engines, massive factories, big blast furnaces, towering smokestacks, large company towns, monopolistic cartels, and vast conurbations. And since iron was the primary raw material used in manufacturing armaments, the coal age ushered in the age of big cannons, massive warships, and total war. "In essence," Mumford concluded, "the entire paleotechnic period was ruled, from beginning to end, by the policy of blood and iron."[3]

Though Mumford made little direct reference to Germany in his excursus on "carboniferous" industrial life, his choice of the Bismarckian phrase "blood and iron" suggests that nineteenth-century Prussia loomed large in his thoughts. The new fossil-based economy impacted the Germans more than other continental Europeans, not least because Prussia controlled the provinces of Westphalia and Rhineland, home to Europe's largest bituminous coal and lignite reserves (known by the somewhat misleading terms "Ruhr" and "Ville" respectively). Ruhr coal provided Bismarck with the military and industrial power he needed in the 1860s to defeat Denmark, Austria, and France in his quest for a unified Germany. The coal-extraction and coal-transport businesses, and the related iron-ore import trade, largely dictated the type of engineering work done on the Lower and Middle Rhine. It was no coincidence that Nobiling's Prussian Navigation Project was undertaken in 1850, at the very moment when Westphalian coal production soared, and that it was completed in 1900, just as Germany emerged as Europe's leading coal, iron, and steel producer. Ruhr coal was the lifeblood of German industry when Bismarck founded the Second Reich (1871–1918); it helped defray post–World War I reparation payments to France and Britain during the Weimar Republic (1919–33); and it stoked Hitler's military-industrial machine during the Third Reich (1933–45).

At first, the new coal mines and iron foundries emerged haphazardly alongside the old commercial and agrarian enterprises in the Ruhr region, filling in the "empty" landscapes and hillsides that lay between the old villages and towns. But as these industries expanded in size and number, they began to take over the riverbanks, usurping the floodplain, and encroaching on farmland and forest. Blue skies and clean air also disappeared as coal-fired factories and belching smokestacks darkened the horizon. Land and air, however, were not the primary "free" resources in high demand. Above all, water was at a premium. Water scarcity itself was never a problem in Rhineland and Westphalia: the Ruhr, Emscher, Lippe, and Erft valleys are all rich in rain-fed runoff, especially during the long winter months. What was increasingly hard to find there was a plentiful and reliable supply of *clean* water, as industrial pollutants accumulated and overwhelmed the self-cleaning capacity of the Rhine's tributaries and feeder streams.

In most respects, environmental problems in Westphalia and Rhineland mirrored those in similar regions around the world. The Ruhr choked in smoke much as did Pittsburgh's Monongahela valley, Ukraine's Donets basin, and Upper Silesia's Könighütte region (now Chorzów, Poland). The Trent, Ouse, Humber, and Mersey rivers of England were all sacrificed in some degree to coal extraction, coal processing, and coal transportation, as were the Ohio, Youghiogheny, and Monongahela in Pennsylvania and the aptly named Coal river in West Virginia. The Ohio was especially overburdened with phenol-laden wastewater due to the large number of coal-coking plants using its basin. One of the leading rationales for "improving" the Danube, Mississippi, Kentucky, and innumerable other European and American rivers was the need to transport ores cheaply. The same rationale lay behind many of the canal-construction projects, most obviously the Bridgewater canal, which linked the coal face in Worsley with the quays in Manchester, and the "anthracite canals," which linked the Pennsylvanian coalfields to Philadelphia and New York.

In some respects, however, Westphalia and Rhineland represented an extreme case of nineteenth-century industrial development, the difference in degree becoming a difference in quality. Great Britain's coal-mining regions were less concentrated than those in Germany, as were its iron and steel foundries. While many British rivers were navigable, none came even remotely close to rivaling the Rhine as a transportation artery, the Thames included (London received much of its coal via sea routes). The Donets basin—located far from the investment-rich and technologically innovative industrial belt of northwestern Europe—was not fully exploited until the Soviet era. Stretches of the Don and Donets were eventually sacrificed to coal production, but the region depended mostly on rail links rather than water transportation. The Ruhr region had everything in one place from the outset: vast quantities of coal, good transportation routes, and ready markets in every direction. It was also ruled by a government willing to translate coal into industrial and military power regardless of the environmental impact. Never before had there been such a concentration of power and pollution in one place, and never since such a clear-cut case of the payoffs and perils of the carboniferous age.

The Bismarckian constitution placed considerable authority over Germany's lakes and streams in the hands of the state governments; and the Weimar constitution later transferred much of that authority to the federal level. Despite this clout "from above," Prusso-German bureaucrats seldom intervened to protect the nation's water quality. It took, for instance, a full three decades of debate before the state legislature finally passed the Prussian Water Act of 1913, a weak piece of legislation notable only because it was the first semblance of a unified state water law. The first Federal Wastewater Act would not come until 1957, and the first truly effective measure would have to wait until 1986.[4]

One reason for the tardiness was that state and national authorities did not want to impose any restrictions that might slow the pace of industrial growth. Another reason was that Germany's atomized past bequeathed an extraordinarily decentralized tradition of water management (a tradition that persists even today). In Bismarck's day, farmers, villagers, industrialists, and urban water suppliers generally depended on private wells and nearby lakes and streams for their water supplies. Water jurisdictions and water laws followed age-old political boundaries, not the contours of watersheds or the dictates of rational planning. There was, of course, friction between upstream and downstream users, but pitched battles were often avoidable simply because supply greatly exceeded demand. Industrial growth, urban sprawl, and transboundary pollution gradually rendered the old laws and practices obsolete, especially in conurbations such as the Ruhr, and by 1913 more rational systems had begun to emerge "from below."

For Ruhr industries and cities, the most important pieces of Prussian water legislation were the Ruhr Dam Cooperative Act of 1891 and the Emscher Cooperative Act of 1904. The first was a special decree that brought the principal Ruhr basin waterworks and hydropower firms together under a single "dam association." The second was a similar decree that brought all water suppliers and water users in the Emscher basin into a single "riparian cooperative." Small steps in themselves, they set a precedent for subsequent cooperatives on the Ruhr, Lippe, Moers, Erft, Wupper, and elsewhere, and soon became the preferred method of water management throughout West-phalia and Rhineland. These cooperatives were privately run organizations with quasi-governmental authority to oversee the apportionment of water to industries, cities, and farmers within their watersheds. The Prussian govern-ment was not directly involved in their management or oversight. In fact, the whole idea behind the cooperative system was to leave water manage-ment to Westphalia's waterworks and industries themselves. The Prussian legislature entered the picture (and reluctantly at that) only to establish the legal foundations for a water-management system that would supersede the old administrative districts, and to ensure that membership would be com-pulsory, thus eliminating the "free rider" problem.[5]

In principle, the riparian cooperatives were designed to handle water-use issues in a rational manner through an institutional framework that was neither overly centralized nor overly decentralized—that is, a frame-work bridging the gap between water management "from above" and "from below." The dam associations, composed as they were primarily of water-works and utilities companies, fulfilled this role tolerably well (though there were human and environmental costs involved with dam construction). The riparian associations—invariably dominated politically and financially by the major industrial enterprises in each basin—did not. In passing the Emscher Act, the Prussian government in effect turned the management of its river basins over to the very mining and metallurgical firms (and later textile and chemical firms) that were causing the most havoc to the environment. Left to their own devices, these firms undertook the whole-sale replumbing of Westphalian and Rhineland rivers to suit their needs— without having to give any thought to river ecology.

Coal and Iron in Westphalia and Rhineland

Prussia in 1815 was economically backward compared with Great Britain, but all indicators pointed toward a rapid development on the Rhine and Ruhr. Westphalia was a right-bank Napoleonic creation that had been one of the most important satellite states within the short-lived Confédéra-tion du Rhin. Rhineland was a left-bank region that had been annexed to France during the revolutionary era. The Congress of Vienna placed both under Prussian control in 1815, primarily to ensure a strong anti-French military presence on the Lower Rhine—a role Prussia ended up playing beyond anyone's wildest dreams. It was known at the time that the Ruhr region possessed some deposits of coal, but far too little was understood about the actual underground wealth of the Lower Rhine catchment area for the Vienna diplomats to realize that they were opening the way for the world's first military-industrial complex. Yet that is what they did. Within a few decades coal and iron rendered the image of "poetically agricultural" Germany—like that of "merry old England"—an icon of the past. The Rhine's future belonged to mechanized production and the noisy steam engine.[6]

State-sponsored exploitation of the Ruhr coalfields began in the mid-eighteenth century when Prussia's Frederick the Great ordered construction of the first transportation dams on the Ruhr, and promulgated a new mining ordinance in Mark and Cleves (Prussia's two tiny possessions before 1815), giving Berlin direct control over Ruhr coal. Coal production, which stood at 70,000 metric tons in 1774, climbed to 333,000 tons in 1814, and 455,000 tons in 1826, the latter figure reflecting Prussia's acquisition of Westphalia and Rhineland. Railroad construction in the 1840s freed the coal industry from dependence on the Ruhr river, facilitating the exploitation of the fields northward toward the Emscher and Lippe basins. Improvements in steam-powered water pumps also made it possible to reach the deeper veins safely. Privatization of the coal mines in 1865, in keeping with the then-fashionable liberal economic policy, further accelerated the upward pace, so that by 1871 the Ruhr region was producing nearly 13 million metric tons of coal annually. Another meteoric rise in production began around 1880 (22.6 million tons) once the shipping lanes on the Lower Rhine were deepened and widened and the railroad network expanded. By 1913, some 115 million tons were extracted annually from the region, a fivefold increase over 1880 and a whopping 1600-fold increase over 1774.[7] Those numbers climbed steadily each year, reaching nearly 130 million metric tons by 1938. Germany's defeat in World War II halted the upsurge, as did the post-1945 switch from coal to oil; but coal production still exceeded 120 million tons annually throughout most of the 1950s before dropping to its current production levels (53 million tons in 1996).[8]

Even today the term "Ruhr" is synonymous with "coal," much as it was in the days of Frederick the Great. But actually the term "Rhine-Ruhr," or "Rhineland-Westphalia" (the name adopted by Germany's behemoth coal syndicate in 1893), is a more apt designation. "Rhine-Ruhr-Emscher-Lippe-Moer" would be even more appropriate, were it not so clumsy, because the coal veins eventually traversed all of these river basins. Each year the coal frontier crept northward, away from the Ruhr and into the Emscher and Lippe basins; new seams were also discovered on the Rhine's left bank, in the Moers district between Krefeld and the Dutch-German border, also well away from the Ruhr catchment area (see figure 4.1). As a barge route, moreover, the Ruhr left much to be desired, despite Frederick's efforts to make it navigable: it was simply too shallow to compete with the Rhine, railroads, and canal networks in the coal-transport business. Coal barges began to disappear from the Ruhr by the mid-nineteenth century, as did the coal harbors, except at Duisburg-Ruhrort, where the Ruhr meets the Rhine. In 1913, of the 18 million metric tons of Rhineland-Westphalian coal shipped, 13 million went by rail and 5 million via the Rhine. None of it was hauled on the Ruhr.[9]

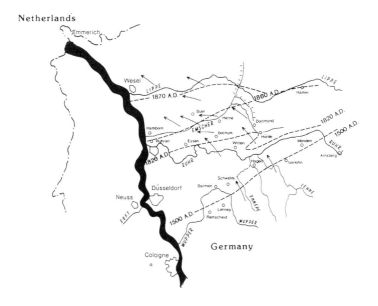

Figure 4.1 The northward growth of Westphalian industry, 1500–1870. The Emscher and Lippe had replaced the Ruhr as the principal centers of production by the 1860s. (Source: Quelle, *Indus-triegeographie*, n.p.)

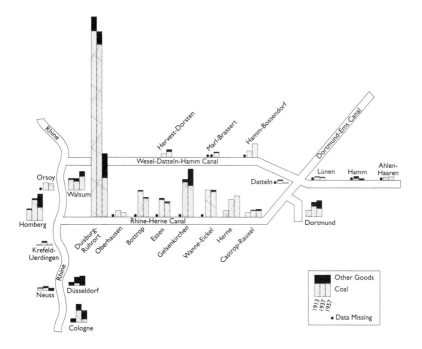

Figure 4.2 Rhineland-Westphalia's coal harbors, 1913–57. The Rhine-Herne canal, known as the "Rhine's extended harbor," was the region's transportation hub. (Source: Schultze-Rhonhof, *Die Verkehrsströme der Kohle*, Abb. 29)

The Ruhr is also synonymous with "iron and steel," but here too "Rhine-Ruhr" or "Rhineland-Westphalia" is more accurate. Before the industrial revolution, Europeans commonly smelted iron ores in specially designed blast furnaces that produced pig iron (cast iron), using a technology first developed on the Lower Rhine in medieval times. Charcoal, made from charred wood, was the main fuel throughout Europe. In the early eighteenth century, British industries began switching from charcoal to coal for iron production as their forests disappeared and the price of imported timber steadily climbed. In 1735, Abraham Darby began using "coke" (coal purified by heating to increase its carbon content) in his blast furnaces. Once perfected, the Darby process spread to the Ruhr and from there across continental Europe, and by the mid-nineteenth century coke had replaced charcoal as the preferred fuel in European metallurgy.[10] Beginning in the

1850s, major technological breakthroughs also came in the steelmaking industries. The Bessemer converter of 1856 made it possible to use coke to heat iron ore and limestone in such a way that the carbon given off by the burning coke went into the iron while the limestone captured the ore's impurities. The result was a purer metal, composed mostly of iron and carbon. A few years later, in 1864, Pierre Martin developed the Siemens-Martin "open hearth" process, which enabled manufacturers to control and vary the carbon content and distribute the carbon uniformly through the molten material. The resulting alloy, when quenched with water, solidified as hard steel.[11]

Until the coal age, metallurgical operations were confined to regions where iron ore could be mined, and where waterpower and fuelwood were plentiful. Around the Rhine this meant the Siegerland, Sauerland, Mark, Eifel, and Saar regions. For Germans the switch from charcoal to coal did not begin until the 1830s. As late as 1825, all 17,000 metric tons of "bar iron" produced in Rhineland and Westphalia were fired with charcoal. But by 1844, nearly two-thirds of the 51,800 tons produced came from coal-driven furnaces. Then, in the 1850s, several coke-fired ironworks appeared in rapid succession: Friedrich-Wilhelm-Hütten (Mülheim), Eintrachthütte (Hochdahl near Düsseldorf), Hütte von Détilleux (Bergeborbeck), and AG Phönix (Ruhrort). In 1848, production of pig iron stood at 71,700 metric tons, only 3.6 percent of which was produced with coke. By 1870, pig iron production had climbed to 766,800 tons, 92.8 percent of which was produced with coke.[12]

Westphalia possessed an abundance of a particular grade of bituminous coal, known colloquially as *Fettkohle*, which lent itself to superior coking and was therefore superbly suited for steel production. Most of this "coking coal" was located in the Emscher river basin north of the Ruhr—yet another factor pushing the seams away from the Ruhr valley. What was lacking in the entire Rhine-Ruhr region, however, was a sufficient quantity of high-grade iron ore to keep the cauldrons producing steadily. There were, to be sure, a few iron deposits in Westphalia, many of them interlayered with the coal veins. But iron was far less plentiful than coal, and by 1870 Rhineland-Westphalia's metallurgical industry was almost wholly dependent on iron ore imported from France, Luxembourg, North Africa, Spain, and (above all) Sweden. Prussia's victory over France in 1870–71, which brought Alsace-Lorraine into the newly created Reich, did little to alleviate the domestic iron ore shortage on the Ruhr, in part because the Lorraine deposits were too meager to sustain the needs of the giant Rhine-Ruhr firms and in part because the iron veins there yielded "minette," a low-grade phosphoric ore ill-suited for steelmaking.[13] (In 1879, the Gilchrist-Thomas method for "dephosphorizing" iron made it possible to use minette in steel production. But it proved more profitable to build a steel industry in Lorraine than to ship the low-grade ore to the Ruhr.)

The ever-mounting dependence on imported iron ore reinforced the trend, already visible in the coal sector, for iron and steel mills to aggregate on the Rhine rather than the Ruhr. It was cheaper to haul iron ore by barge than by rail, and the Rhine offered much better access to the world's iron suppliers. In addition, each improvement in manufacturing tended to reduce the amount of coal needed in iron production. Whereas in 1850 it took ten tons of coal to produce one ton of iron, by the end of the century a single ton of coal could produce two to three tons of iron. For iron and steel manufacturers this meant that it was increasingly economical to locate as closely as possible to their iron ore supplies. Thyssen moved from the Ruhr to the Rhine in 1895, Krupp soon thereafter, and within a decade most other iron and steel firms had cut their ties to the Ruhr as well.[14]

By 1913, Westphalian and Rhineland industries were importing some 10.5 million metric tons of iron ore annually from overseas, more than 70 percent (7.5 million metric tons) coming by water transport. Nearly all of the imported ore came up the Rhine through Rotterdam. Dependence on a Dutch harbor worried the Germans enough that they decided to construct the Dortmund-Ems canal. Finished in 1899, it created a more direct route from Sweden to Westphalia by way of the Emden port and Ems river. From the outset, however, it was too small to compete effectively with Rotterdam and the Rhine and thus served only as a backup route. In 1914, moreover, the Rhine-Herne canal, which traversed the Emscher basin and connected to the Dortmund-Ems canal, became operational. Dubbed the "Rhine's extended harbor," Rhine-Herne cemented once and for all the Rhine's preeminence in the Rhineland-Westphalian iron and steel industry (see fig. 4.2). In 1895 only about one-quarter of the region's total ouput was produced on the Rhine; by 1914 that share had climbed to over 60 percent.[15]

Water Shortages

Coal and iron transformed the Rhineland-Westphalian region from an agrarian to an industrial landscape in a remarkably short time. By 1900, mines, factories, and smokestacks filled the space where only a few decades earlier trees, birds, orchards, grains, and root crops (principally potatoes) had reigned. These industrial transformations had a profound impact on the region's river and stream hydrology. Each steam engine required 160 to 315 liters of water per unit of horsepower. Each ton of pig iron required some 7000 to 12,000 liters of water. Each ton of coal took 1250 to 1800 liters of water for cleaning and other purposes. Coking and other processing could bring that total up to 3000 liters of water per ton of coal.[16] The Krupp steel works in Essen alone used as much water each day as a typical German city of 250,000 to 300,000 inhabitants.[17] Since much of this industrial water was utilized for cooling and cleansing purposes, it could be (and often was) recycled. But vast quantities of water were also consumed and not replaced, and as the number and size of the industries climbed, so did Westphalia's water deficit. By the end of the nineteenth century water shortages had become a persistent source of concern. Equally vexing was the problem of waste removal.

The Ruhr's mining and metallurgical firms were more than powerful enough to reengineer the watersheds of Westphalia and Rhineland to suit their needs. And they did not hesitate to do so, especially once the Emscher Act allowed them to ignore the needs of local fishermen and private well owners. Their first challenge was how to distribute the waters of the Emscher, Ruhr, and Lippe (Westphalia's three major Rhine tributaries) to handle most efficiently the effluents from mining and metal processing. Their solution—at first unofficial and later official—was to utilize the Emscher as an open-air sewer (or *cloaca maxima* as it was dubbed); to preserve the Ruhr as the main source of clean water for industry; and to designate the Lippe as the supply source for the coal canal network (the Lippe Lateral, Dortmund-Ems, and Mitteland canals). The second challenge was to create a safe and reliable water supply that could meet industry's needs even during the worst drought years. Large dams—based on technology developed in the United States and Great Britain in the 1890s—provided the answer to this dilemma.

The Emscher river was without doubt the first and worst affected by this industrial plumbing project. Its short length (109 kilometers) and small watershed (850 square kilometers) made it less valuable as a reliable source of *fresh* water for industry than the Ruhr and Lippe to the south and north. It snaked, moreover, through the middle of the Westphalian coalfields, making it an irresistible target for the dumping of industrial effluents. It earned a reputation as Westphalia's sewer in the early days of coal mining, and the dirtier it got the more official its status became. By 1903 the Emscher was receiving wastewater from 1.5 million urban residents, 150 coal mines, and 100 other factories each year.[18] By 1910 only about half of its annual water budget was coming from natural runoff; the remainder, 96 million cubic meters, came from industrial (89 percent) and urban (11 percent) discharges.[19] The coal industry alone accounted for most of the effluents. Westphalia's bituminous mines tended to be extraordinarily deep, and water pumps were needed to keep the tunnels from flooding. On aver-

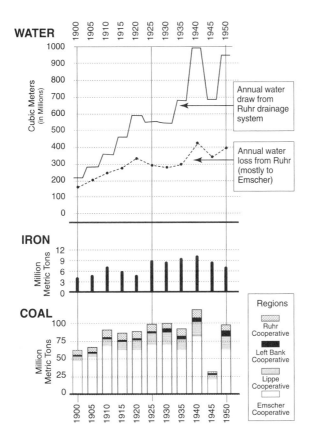

Figure 4.3 The coal, iron, and water resources of Rhineland-Westphalia, 1900–1950. The needs of mining and metallurgy largely dictated water-use policies on the Lower Rhine. (Adapted from Ramshorn, "Die Wasserwirtschaft," *Glückauf,* 9)

age, 1700 liters of "mine water" were pumped out of the mine for every ton of coal extracted, much of it laden with chlorides. The mine water was utilized to wash the coal and then was dumped (along with the chlorides and coal dust) into the Emscher untreated.[20] The coking process produced an additional waste product—phenol—which lent the Emscher valley its telltale stench.[21]

Unfortunately, the Emscher was poorly suited to its designated role as an industrial sewer. Its natural gradient was slight and its water level highly susceptible to seasonal variations. It meandered slowly all over its valley. Its waters stagnated easily in the marshlands, making it a good host for typhoid, cholera, and other waterborne diseases. The addition of industrial effluents nearly doubled its annual flow without appreciably accelerating its velocity, creating even more problems with stagnant water than before. The nearby coal mines, meanwhile, caused subsidences in the landscape, especially in the urban areas between Herne and Oberhausen, adding to the number of stagnant and foul-smelling pools. Chloride, coal dust, and phenol overwhelmed the stream's self-cleaning capacity, killing vegetation

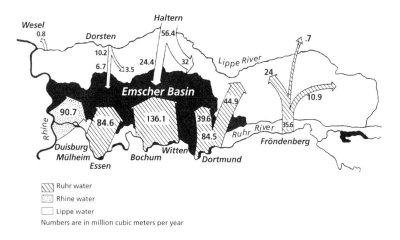

Figure 4.4 Water transfers between the Rhine, Ruhr, Lippe, and Emscher circa 1950. Most of the transfers flowed from the Ruhr and Lippe into the Emscher basin. (Adapted from Ramshorn, "Die Wasserwirtschaft," *Glückauf*, 11).

and fish and clogging the streambed; in some places the salt content was so high that it damaged the nearby agricultural fields. Not surprisingly, disease rates were higher on the Emscher than elsewhere in Germany. Sent to investigate a typhoid epidemic that struck some three thousand people in Gelsenkirchen in 1901, killing over three hundred, Rudolph Emmerich filed this report on water conditions on the Emscher: "Here we find a black, thick, swampy, rotten and fermenting manure, which hardly moves: during summer, gas bubbles burst, poisoning the surrounding area." He noted further: "The fermentation and putrification of these enormous quantities of disgustingly dirty and muddy waste water are intensified to the highest degree by the hot water from innumerable steam engines; as a consequence, the waste water often reaches breeding temperature. The waste water to be found in Gelsenkirchen is the most disgusting in the world."[22] Within a few short decades the Rhineland-Westphalian coal industry had turned the once clear and free-flowing Emscher into, as one Prussian parliamentarian put it in the 1920s, "the river of hell."[23]

If the Emscher was hell, the Ruhr was purgatory. It had a much steeper natural gradient and a much larger watershed (4500 square kilometers) than the Emscher, its swifter current lending it a much stronger self-cleaning capacity. Pollution was something of a problem on the Lower Ruhr stretch, especially between the cities of Mülheim and Duisburg-Ruhrort, where paper mills and metallurgical foundries cluttered its banks. In general, however, Ruhr water was safe and clean—not safe enough for humans to drink unpurified, but certainly clean enough for industries to use as a freshwater source. In fact, the idea behind dumping wastewater into the Emscher was to keep the Ruhr's waters as clean as possible. But serious environmental problems arose precisely because of this artificial hydrology: factories pumped millions and millions of cubic meters of water out of the Ruhr each year, but returned almost none of it to its bed (since it was either consumed or dumped as wastewater into the Emscher). In 1900, the Ruhr's water deficit was 150 million cubic meters. By 1905 it had climbed to 175 million, and by 1910 to 200 million.[24] Less water in the Ruhr meant less self-cleaning capacity and less velocity to carry away the debris caused by riverfront cities and riparian mills. It also meant serious shortfalls of clean water for the city waterworks.

The ever-mounting Ruhr deficit created water management headaches even in wet years, especially during the rainless summer months, when the river typically ran low. When drought conditions prevailed, Ruhr industries and cities found themselves in truly desperate straits. August Thienemann, commissioned to investigate the impact of the 1911 drought on the Lower Ruhr, concluded that the "liquid substance" flowing in the riverbed did not deserve to be called water. He offered various alternative names: "yellow soup with foam," "milky gray mud," "oil slick." The Ruhr Fisheries Association, he noted, had pulled two buckets of test water from the Ruhr, one taken just upstream and one just downstream from the gigantic Friedrich-Wilhelm-Hütte in Mülheim. Fish placed in the upstream bucket managed to survive, but those in the downstream bucket died almost instantly of potassium cyanide poisoning. "Here the Ruhr is really nothing but sewage," Thienemann concluded. "Twenty-eight degrees Celsius. A brown-black broth that reeks of prussic acid and is completely dead."[25]

Such extreme conditions on the Lower Ruhr occurred only during the rare occasions when a summer dry spell followed a winter drought. In normal years, the river's self-cleansing mechanisms kept the water relatively free of pollutants. But with the demand for industrial water rising every year, summer shortages were becoming the rule rather than the exception, and the intermittent droughts served to underscore just how fragile Westphalia's water supply had become. Legal disputes between upstream and downstream factories, and between industries and cities, began clogging the court systems, lending a note of urgency to the quest for a solution. With "water wars" looming, pressure for reform came from regional authorities, industries, scientists, and, above all, the public and private waterworks charged with the task of providing potable water to city residents. The key to success, as it turned out, lay in cooperation, or at least in the creation of "cooperatives."

Dam Associations and Water Cooperatives

A giant step toward improving Westphalia's water supply was taken in 1899 with the creation of the Ruhr Dam Association (Ruhrtalsperrenverein), a few years after the passage of the Ruhr Dam Act of 1891. Headquartered in Essen, it was a consortium of the waterworks and power plants most affected by the river's periodic water shortages. According to its founding charter, the chief aim was "to improve the quantity and quality of the water supply on the Ruhr through the construction of dams in the Ruhr watershed."[26] The driving force behind the Association was Otto Intze. A professor at the Technological University of Aachen, Intze was Germany's best-known champion of large dams. He had earned his reputation as a dam builder in the early 1890s, when he designed city water reservoirs similar to those then being pioneered in Great Britain and the United States. His first dam, the Eschbach, was commissioned by the town of Remscheid; completed in 1891, it held 1.7 million cubic meters of water. He designed similar dams for Ennepetal, Lüdenscheid, Marienheide, Wuppertal, and other cities. Then, in 1897, he turned his considerable talent and energy to convincing Westphalia's industrial and urban planners that dams were the panacea for the Ruhr's water woes. His reputation and energy helped ensure that the Association was up and running within two years, and he himself was commissioned to design its first dams.[27]

Over the next seventy years, the Ruhr Dam Association constructed enough dams to hold 469 million cubic meters of water: Lister (1912: 21.6 million), Möhne (1913: 134.5 million), Sorpe (1935: 70 million), Verse (1952: 32.8 million), Henne (1955: 38.4 million), and Bigge (1965: 171.7 million).[28] The Möhne reservoir alone submerged over 12 square kilometers of farm and meadow land; six villages were affected (Kettlersteich, Delecke, Drüggelte, Körbecke, Stockum, and Wavel), 200 public and private buildings flooded, and more than 700 people displaced.[29] The environmental and human costs were acceptable to most Germans at the time because the dams promised to satisfy Westphalia's most immediate and pressing industrial need: bountiful amounts of clean water. Dams enabled local authorities to avoid pitched legal battles between industries and waterworks, and to supply the Ruhr streambed with sufficient water throughout the year, all but eliminating the low-water conditions that once plagued the river (though as late as September 1929 emergency pumps were used to siphon Rhine water into the Ruhr when it went completely dry).[30]

By the 1930s, the Ruhr was so rich in dam-fed water that it could regularly supply 500 to 700 million cubic meters annually to industries and cities—and even as much as one billion cubic meters in the early years of World War II (see figs. 4.3 and 4.4). By way of comparison: all of Germany's waterworks combined drew around 2.5 billion cubic meters of water annually from Germany's rivers and groundwater supplies during the 1930s. In other words, the Ruhr valley alone provided about one-fourth of Germany's entire annual freshwater supply—supreme testimony to Westphalia's unquenchability.[31] (Small wonder that British and American bombers targeted the Ruhr dams so often, despite the difficulties of striking them with the then-available technology; had they been destroyed, Hitler's war machine would have died of thirst.)

The superabundance of clean water, however, did not resolve the more knotty problem of how to dispose of the ever-increasing amounts of industrial and human wastewater generated by Westphalian industries and cities. Ruhr water, once removed from the river, did not stay clean for long: it flowed through coal-processing plants, iron and steel foundries, pulp-and-paper mills, textile factories, city toilets, and irrigation pipes. More water meant more industrial and urban growth. More industries and cities meant more effluents. And more effluents meant more stench in the densely settled Emscher valley.

The handling of Westphalia's industrial and urban wastewater was pioneered by a second association, the Emscher Cooperative (Emschergenossenschaft), also headquartered in Essen. Founded in 1899, the same year as the Ruhr Dam Association, it did not become operational until five years later when Prussia passed the Emscher Act. The cooperative counted among its members virtually all the corporate entities that utilized the Emscher and its feeder streams for waste disposal, as well as the urban communities that once relied on the river for drinking water. In principle, the goal of the Emscher Cooperative was to coordinate water policy between water suppliers and water users throughout the Emscher's catchment area. In practice, the polluting industries kept the cooperative focused on the more narrow task of waste disposal. Water management on the cheap was the cooperative's principal concern: protection of the Emscher as an ecological habitat was not even remotely on the agenda.

Predictably, the cooperative found that the least expensive disposal method was to channel the effluents as swiftly as possible down the Emscher and into the Rhine. An investigation conducted by its own scientists between 1906 and 1912 convinced the cooperative's directors that Rhine dumping was a viable long-term solution. The researchers discovered, to be sure, "a noticeable increase in mineral components and suspended matter on the right bank of the Rhine" just below the Emscher mouth. They also noted a significant rise in chloride, lime, and magnesium levels directly on the Emscher mouth. Water samples, for instance, taken from the Rhine in November 1906 indicated a chloride content of 68.0 milligrams per liter above the Emscher mouth, and 176.0 milligrams per liter below, a threefold increase. But samples taken farther downstream indicated that chloride and other contaminants dissipated quickly once they mixed with Rhine waters. "The Rhine is still capable of absorbing far more pollution than it now receives from the Emscher without any danger," was the cooperative's self-serving conclusion.[32]

The cooperative gave top priority to reengineering the Emscher so that it would more closely resemble the open-air sewer it had already become. Its length was shortened from 109 to 81.5 kilometers, which gave it a steeper gradient and therefore a faster flow. Its bank and bed (and that of its entire tributary system, some 300 kilometers in length) were lined with cement to prevent seepage and to counteract the land depressions induced by underground mining. In polders (land areas that lie below the water table), the cooperative installed special pumps and drainage systems so that stagnant water could be routinely returned to the Emscher or one of its feeder streams. The original Emscher mouth at Duisburg-Beek was moved 2.8 kilometers north in 1910, and then 6.3 kilometers farther north in 1949, both times in an effort to overcome the persistent problem of land subsidences from underground mining. When these efforts failed, the cooperative decided to abandon any pretense of a natural-looking river. Since 1951, the Emscher "mouth" has consisted of nothing more than two gigantic steel pipes, each 2.2 meters in diameter, extending deep into the middle of the Rhine.[33]

The cooperative's second priority was the installation of wastewater and sewage treatment plants. Here the influence of Karl Imhoff was decisive. Imhoff was the cooperative's leading sanitation engineer and also the inventor of the "Emscher well," better known outside Germany as the "Imhoff tank." It was a primitive purification system, basically consisting of two tanks for collecting and siphoning sludge, not all that innovative even by the standards of the day. But it was attractive to the cooperative because it promised to reduce much of the river's stench at a minimal cost. The first systems began to operate in 1907, and by 1914 a total of 138 Imhoff tanks were in operation at 23 facilities. Although Imhoff marketed his tanks as a mechanical and biological cleansing combination, most of the early versions were in practice mechanical. Wastewater was fed into a holding tank and the sediment allowed to settle to the bottom. The sludge was then dried and sold as a low-grade nitrogen fertilizer (heavy metals and all), while the "purified" water was piped into the Emscher or one of its tributaries. A more sophisticated and effective biological treatment process (the so-called trickle method) was also available, but the cooperative, ever watchful of its budget, managed to postpone its widespread use until after 1918.[34] Technological improvements continued thereafter, but the practice of getting by as cheaply as possible remained the guiding principle.

A second riparian association, the Cooperative for a Clean Ruhr (Verband zur Reinhaltung der Ruhr), was founded in 1911. It began as a branch of the Emscher Cooperative, then became a self-administered entity shortly after World War I. In principle, the two cooperatives had diametrically opposite tasks: the Emscher group existed to get rid of wastewater, the Ruhr group to safeguard a freshwater supply. But from an environmental perspective, the difference was insignificant. Faced with the task of removing metallurgical pollutants from the Lenne tributary, for instance, the Ruhr Cooperative decided that the simplest solution was to construct a small dam on the Ruhr just below the Lenne mouth. The dam raised the Ruhr's water level by nearly five meters, slowing the river's flow and creating a backwater. The backwater served as an in-river holding tank, allowing the metallic tailings to settle as the rest of the Lenne water flowed into the Ruhr—a removal method that was as economically cost effective as it was environmentally shortsighted.[35]

In other respects, too, the Ruhr Cooperative favored short-term results over long-term solutions. One of its first major projects, for instance, was the construction of an industrial sewage canal on the Lower Ruhr running from Mülheim to the Rhine. Completed in 1924, the canal functioned as a kind of ersatz Emscher for the Lower Ruhr, absorbing the effluents from the urban-industrial centers of Mülheim, Oberhausen, and Duisburg that had previously been dumped into the Lower Ruhr. As with the Emscher, the ultimate destination for the debris was the Rhine. "The burden on the Rhine as regards wastewater is small," explained the ubiquitous Imhoff, now the Managing Director of the Ruhr Cooperative, in 1928. "Except to protect the fisheries, there is no rationale whatsoever to do anything about the pouring of wastewater in the Rhine because its water volume is so large that it can absorb and cleanse the waste itself." He added: "Wastewater, which would be in every respect dangerous and even poisonous in the Ruhr, is rendered harmless when poured into the middle of the Rhine."[36]

Actually, matters were not quite as simple as Imhoff suggested. One of the most intractable problems that bedeviled both cooperatives during their first decades was the removal of phenol from Emscher and Ruhr wastewater. Phenol was a hazardous byproduct of the coking and coal-tar industries. It had some industrial applications, but most firms considered it a waste

COLOGNE

The real Eau de Cologne, and its effect upon the noses of three illustrious individuals

A caricature of Rhine Romanticism in 1855. Cologne was among
the first Rhine cities to experience industrial pollution. (From Doyle,
The Foreign Tour, p. 7)

A View on the Rhine

A caricature of Rhine tourism in 1855. (From Doyle, *The Foreign Tour*, p. 14)

"An argument between Father Rhine and a River Engineering Authority." Sketch by Jakob Albrecht, 1861. (From *Der Alpenrhein und seine Regulierung*, p. 103)

"The Tamed Rhine." Sketch by Hedwig Scherrer. (From *Der Alpen-rhein und seine Regulierung*, p. 235)

LORELEY, SALMFANG

F. NITZSCHE

Rhine postcard kitsch, c. 1900. The salmon industry was already
in decline, but the image of the lone Lorelei fisherman was alive and
well. (Rhein-Museum, Koblenz)

Reederei
‹BRAUNKOHLE›
G · M · B · H
WESSELING

SCHIFFAHRT · SPEDITION · UMSCHLAG
LAGERUNG · BOOTEKOHLENHANDEL
TRANSPORTVERSICHERUNG

Advertising for a Rhine transport firm, c. 1950. The ship's belching smokestacks are foregrounded, while Cologne's famous cathedral forms part of the hazy backdrop. (From Wasser- und Schiffahrts-direktion Duisburg, *Der Rhein: Ausbau*, p. 401)

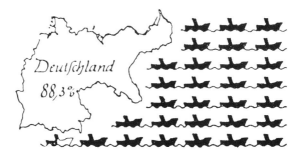

A graph from the Nazi era. Pre-World War I dye production in
Germany, Switzerland, and Great Britain is depicted as a naval race.
(From Andersen and Spelsberg, eds., *Das Blaue Wunder*, p. 19)

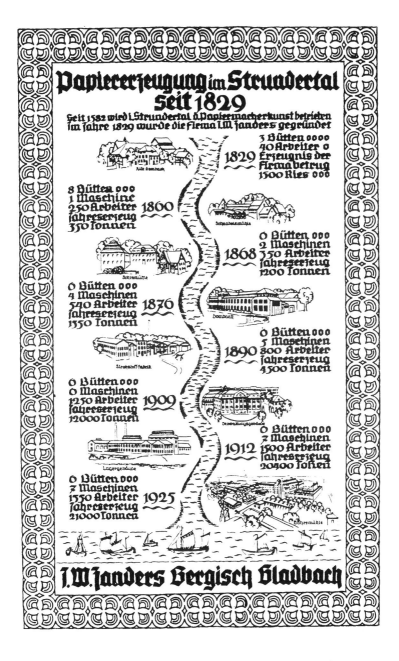

An advertisement for the J. W. Zanders Paper Company. Zanders built eight paper mills on the small Strunde river between 1829 and 1925. (Rhein-Museum, Koblenz)

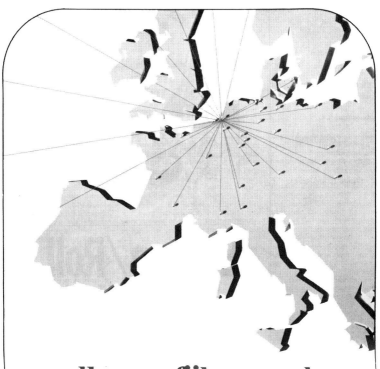

alle wege führen nach
rotterdam·europoort

Hier sind die geographischen Vorzüge mit gewaltigen technischen Mitteln zum Aufbau des grössten Verteilerzentrums in Europa genutzt worden.

Die geographischen Vorzüge: Günstige Lage an der Rheinmündung. Ganz in der Nähe der industriellen Ballungsgebiete. Angeschlossen an das europäische Verkehrsnetz durch Wasserstrassen, Autobahnen und Schienenwege. Auf dem Seewege verbunden mit Nachbarländern und Handelspartnern in Übersee.

Die technischen Voraussetzungen entsprechen schon heute den Anforderungen von morgen. Hafenbecken für Superschiffe. Modernste Anlagen und Geräte für den schnellen Umschlag von Stückgut, Mas-sengut und Flüssiggut. Container-Terminals, die zu den bestausgerüsteten und modernsten der Welt zählen.

Hinzu kommt: Ein modernes flexibles Zollsystem, das die Güterabfertigung noch erheblich beschleunigt. Fazit: Mit über 30.000 einlaufenden Seeschiffen und 250.000 Binnenschiffen sowie einem Güterumschlag von za. 220 Million Tonnen im Jahr ist Rotterdam Europoort heute der am stärksten frequentierteste Hafen der Welt.

Was das für die verladende Wirtschaft bedeutet? Ob Export oder Import - Gütertransporte via Rotterdam helfen Zeit und Geld sparen. Deshalb führt kein Weg an Rotterdam vorbei. Alle Wege führen heute nach Rotterdam Europoort.

rotterdam·europoort

Auskünfte: Städt. Hafenbetrieb - Stieltjesstraat 27 - Rotterdam 3020

An advertisement celebrating Rotterdam's status as the world's busiest harbor. The caption reads, "All Roads Lead to Rotterdam-Europoort." (*Der Hafenkurier*, February 1972, p. 77) product and simply dumped it into nearby streams. Most of the phenol (around 6000 metric tons each year) found its way to the Emscher, the rest to the Ruhr. The Imhoff tanks were ineffective at removing phenol, so most of the phenol eventually flowed untreated into the Rhine.[37] Flushing a problem downstream was not a strategy that normally caused any consternation for the Emscher and Ruhr cooperatives; it was, in fact, their preferred method of waste removal. But phenol was more than just a waste product: it was a public relations nightmare for the coal industry. Phenol had an easily recognizable odor, one that aroused the ire of riverbank residents and put regulatory agencies on the scent. Salmon, eel, and other "fatty" fish absorbed phenol into their bodies as they swam through contaminated streams; if caught by fishermen before they had a chance to shed the poison, the fish stank and were inedible.

As the complaints grew louder, it became increasingly difficult to justify the cooperatives' practice of dumping untreated phenol into the Emscher and Ruhr, especially when such dumping meant that the phenol would just flow into the Rhine, or worse (as happened in Essen in 1925) directly into the urban water supplies of Westphalia.[38] Fortunately, by 1926 a new technology made it relatively easy to remove phenol from industrial wastewater. But these dephenolization plants were costly to construct, and the market value of the recovered ammonia and carbolic acid was low. There was, therefore, little incentive for the cooperatives to build them except as a sop to public opinion. Only three such plants had been built on the Emscher by 1929, when the Great Depression temporarily brought a halt to further construction. Only nine had been built by 1939, when World War II again delayed further construction. As late as 1956 there were still only twenty dephenolization plants, barely enough to remove half the Emscher's phenol content (see figure 4.5).[39]

The situation was similar in the Ruhr basin, even though the avowed goal of the Ruhr Cooperative was to keep the river clean. The cooperative ascertained in 1925 that virtually all of the Ruhr's phenol effluents originated from a single tributary, the Oelbach, where five coal mines and one coal-tar distillation factory were clustered.[40] Even though the problem could have been solved easily, the cooperative was slower than its Emscher counterpart in constructing the needed dephenolization plants. "The complaints of the fishermen are exaggerated," Imhoff noted with characteristic bluntness, "and, from an economic viewpoint, utterly meaningless."[41]

Two other water associations—the Lippe Cooperative and the Left

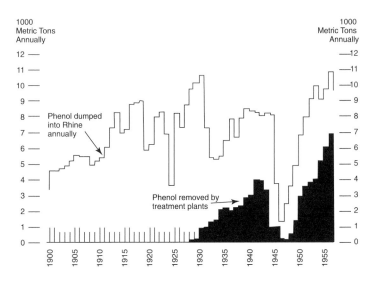

Figure 4.5 The Emscher's annual phenol load, 1900–1950.
The Emscher was the principal source of phenol contaminants in
the Rhine into the 1950s. (Source: Ramshorn, *Die Emscherge-nossenschaft*, 51)

Rhine Drainage Cooperative—deserve brief mention. The Lippe river is
230 kilometers long with its mouth into the Rhine at Wesel. Its catchment
area of 5000 square kilometers lies directly north of the Emscher and Ruhr
basins. Once used for fishing and agriculture, the river became increasingly
polluted with the northward advance of the coalfields, and in 1926 the principal water suppliers and users joined together. The Lippe Cooperative was
a hybrid of the other associations. Its principal task was to ensure that the
Lippe, Dortmund-Ems, and Mitteland canals were supplied with sufficient
water for year-round navigation. The construction of dams in the upper
reaches of the Lippe basin, following the model of the Ruhr Dam Association, solved this problem. Its second task was to keep the water in the upper
stretches of the river (roughly the catchment area upstream from Hamm)
clean and safe for agricultural and industrial use. Here the Ruhr Cooperative served as the model. Its third task was to channel the coal effluents in
the lower stretches of the river into the Rhine cheaply and efficiently. Here
the Emscher Cooperative provided the model.[42]

The Left Rhine Drainage Cooperative, headquartered in Moers, was the only left-bank cooperative in the Ruhr bituminous fields and thus the only one of its kind in Rhineland. It was founded in 1913, but did not become fully operational until the mid-1920s, by which time hydrological conditions in the region resembled those on the Emscher a half century earlier. Since the region was bereft of any sizable rivers, coal firms just dumped their mine water into the various rivulets and streams (most notably the Moers), ruining them all in turn. Given the prevailing conditions, the Left Rhine Cooperative adopted the same drainage methods used on the Emscher: it shortened, straightened, and canalized the region's small streams and brooks, turning them into efficient drainage ditches; and it added polder pumps as needed. Most of the industrial waste was channeled eastward to the Rhine, the rest westward to the Meuse (and from there to the joint Rhine-Meuse delta in the Netherlands).[43]

Once mining operations ensconced themselves on both sides of the Rhine, the coal firms soon started clamoring for the right to dig directly under and around the Rhine channel itself, where an estimated one billion metric tons of bituminous lay buried. Any digging in the Lower Rhine valley, however, raised the specter of land subsidences, with serious implications for the river's navigability. A test seam was dug in 1922 in the vicinity of Duisburg, but vociferous opposition from the Rhine Commission and Duisburg harbor officials managed to halt any further excavation. Then in 1937, Nazi leaders—keen on using coal-gasification techniques to produce gasoline and aviation fuel and worried about possible coal shortages in the event of war—gave the green light to full-scale exploitation. Between 1937 and 1950 (when the new Federal Republic of Germany halted the practice) about 42 million metric tons of coal were extracted from the Lower Rhine valley, mostly along the 15-kilometer stretch between Duisburg-Hochfeld (kilometer 775) and Walsum (kilometer 790). The result was the partial "Emscherization" of the Lower Rhine: a sharp drop in the surrounding water table, a vast increase in erosion and other channel problems, and a dramatic proliferation of polders as the exhausted seams collapsed—all necessitating a costly re-engineering of the river's banks and bed in the Duisburg area.[44]

Ultimately, the dam and riparian cooperatives of the Rhine-Ruhr regions served only one king: coal. Behind the facade of rational water management lay the crude calculations of the industrial mindset. The Ruhr's mountains and valleys were but water-storage sites. The Emscher, Lippe, Ruhr, and

Moers were but pipes and sewers. The Rhine was but a sink and toilet. The cooperatives wiped the grime and dirt off the face of their king and made it "disappear" down the Rhine—all in a shortsighted attempt to avoid facing the long-term environmental impact of the carboniferous age.

The North Rhine Lignite Field

The Erft river, a left-bank Rhine tributary that flows through the North Rhine Lignite Field, also faced Emscher-like water woes. Formed in the Miocene epoch, Rhine lignite lies buried beneath layers of riverine sand and clay in a triangular area about 2500 square kilometers in size between Cologne, Mönchengladbach, and Aachen. Geologic activity tilted and fractured what was once a single long seam into several distinct deposits and fault blocks. The Erft (117 kilometers long with a catchment area of 2040 square kilometers) uses one of these fault lines as its channel bed.

Open-pit mining began in the early nineteenth century in the shallow "Ville" deposits just west of Cologne. Long the center of the region's mining operations, "Ville" still stands metonymically for all Rhine lignite, much as "Ruhr" stands for all Westphalian coal. But after 1945, when new extraction machinery rendered it feasible to do open-pit mining hundreds of meters below the surface, firms began to migrate westward in search of deeper and thicker seams, and today's mines are all west of the Erft, where the lignite was once too deep to reach.[45]

Open-pit mining made raw lignite much cheaper to extract than bituminous, but its high water content and low carbon value rendered it unsuitable for iron and steel production, and unprofitable to transport over long distances. Unable to compete with the Ruhr fields, the Ville languished for much of the nineteenth century, providing cheap fuel for local pottery, brickmaking, and liming operations, but not much else. Then in 1877 a new technology enabled entrepreneurs to press most of the moisture out of raw lignite. The resulting "briquettes," though still inferior to bituminous coal, had sufficient carbon content to use as a fuel for generating electricity and for home heating. Soon thereafter, the North Rhine Field emerged as the hub of Rhineland-Westphalia's briquette-making and electrical energy industries. Lignite production, which stood at a mere 128,000 metric tons in 1880, climbed to 20.3 million in 1913, 54.9 million in 1937, and 83.4 million in 1957. Power lines began to crisscross the Ville landscape, as public and private utilities built power plants near the mines to take advantage of the cheap energy source. By 1913, Ville utilities were producing 33 million kilowatt hours of energy. Production zoomed to 526 million kilowatt hours in 1937, and 2.6 billion 1957.[46] Briquette making began to decline in the 1960s, a victim of new lignite-processing technologies and competition from natural gas and other home heating fuels. But power plants and power lines continue to dominate the Ville horizon: of the 101.4 million metric tons of lignite extracted from the North Rhine Field in 1994, over 80 percent was used for generating electricity.[47]

Since the 1950s, the lignite fields have been controlled by Europe's largest energy consortium, Rheinische-Westfälische Elektrizitätswerk (RWE), headquartered in Essen; and its subsidiary, Rheinische Braunkohlenwerke (or "Rheinbraun"), headquartered in Cologne. (The political backdrop for this consolidation was the extinction of Prussia in 1945 and the establishment of North Rhine–Westphalia as a state within the Federal Republic.)[48] Consolidation of the lignite and utilities industries allowed the mines to operate on a scale unheard of a century earlier. Today's open-pit mines cover nearly 100 square kilometers of land. Exhausted mines account for another 150 square kilometers. Bucket-wheel excavator machines extract up to 240,000 metric tons of lignite in a single day, twice as much as was pulled from the Ville in all of 1880. The lignite is processed on the spot and then fed directly into one of RWE's five mammoth power stations, which transmit electricity all over Europe—72,100 gigawatts in 1993 alone.[49]

Mining on such an imperial scale brought with it king-size environmental headaches, especially in the Erft basin, but also in the Inden-Rur valley, which is part of the Meuse watershed. Most visible to the eye were changes in the topography, especially the scarring caused by topsoil and gravel removal, the new hills composed of coal dust and other residue, and the artificial lakes created by ground sinkings and abandoned mines. Companies that can move mountains can also move rivers and people. The Erft's channel has been diverted several times to make way for the pits, and the Erft canal has been built to ease discharges into the Rhine at Neuß.[50] Since 1952, over fifty towns and villages have been moved, and nearly 30,000 humans displaced, to make room for new mines and power plants, including the towns of Bottenbroich, Berrenrath, Mödrath, Grefrath, Habbelrath, Morken-Harff, Königshoven, Lich-Steinstraß, and Garzweiler.[51]

Most burdensome to mine operators, and to those who live in the North Rhine Field, has been the long-term impact of mining on the local hydrology. Open-pit mines cut deep into the earth, sometimes (as in the case of the Fortuna mine near Bergheim) several hundred meters deep, well below the water table. By the 1950s, the Ville's water table had dropped by more than 70 meters, enough to undermine the local woodland ecology, reduce agricultural productivity, and render private and public water wells inoperable.[52] To keep the mines from being flooded with groundwater, companies have had to install powerful pumps to siphon off the seepage (along with coal contaminants) and dump it into the Erft—with the Rhine, of course, as the ultimate destination. In 1960 alone, 957 million cubic meters of water were pumped in order to extract 79.8 million metric tons of lignite. Since these amounts far exceed the Erft's carrying capacity, the Cologne rim canal, a special industrial conduit, was constructed in 1957 to collect a portion of the mine water and other effluents and dump it into the Rhine near Worringen.[53]

Not surprisingly, the lignite-energy industry has been the single biggest consumer and polluter of the Erft basin's water for the past century. During the 1950s, for instance, the gigantic RWE Kierdorf alone extracted nearly 25 million cubic meters annually. Other power plants—along with leather, beer, sugar, paper, and textile factories—brought the annual total up to around 65 million cubic meters each year.[54] Worst hit by pollution was the 50-kilometer stretch between the Ville mines and the Erft mouth, where most of the briquette factories were located; by the 1950s, they were pumping over 30 million cubic meters of coal-contaminated wastewater into the Erft. So much water was going in and out of the Erft via pumps, pipes, and canals that there was almost nothing left of the valley's natural hydrology except for a few scattered remnants of original streambed.[55] Rhineland industries were as innovative as their Westphalian counterparts in replumbing the surrounding watersheds for the purpose of securing a reliable supply of fresh water. The Eifel-Rur Dam Association (Talsperrenverband Eifel-Rur), based in Aachen, built a series of dams, mostly in the rainfall-rich North Eifel: Urft (1906: 45.5 million cubic meters of water), Obermaubach (1934: 1.65 million cubic meters), Heimbach (1935: 1.21 million), Obersee (1959: 20.8 million), Rur (1959: 181.8 million), Olef (1961: 19.3 million), and Wehebach (1981: 25 million).[56] The Association's first dam was designed by Intze, the same engineer who oversaw construction of the early Ruhr dams.

As regards wastewater disposal, however, the Rhineland lagged far behind Westphalia. There was no riparian association in the North Rhine Field like those on the Emscher and Ruhr, at least until the Erft Cooperative (Erftverband) was belatedly established in 1958, a half century later than its counterparts on the Rhine's right bank.[57] As a consequence, the Erft became one of the most overutilized and unregulated rivers in the Rhine's entire catchment area—yet another despoiled artifact of the carboniferous age.

"To an American conservationist, one of the most insistent impressions received from travel in Germany is the lack of wildness in the German landscape," Aldo Leopold wrote, after returning from a three-month research visit in 1935:

> Forests are there—interminable miles of them, spires of spruce on the skyline, glowering thickets in ravines, and many a quick glimpse "where the yellow pines are marching straight and stalwart up the hillside where they gather on the crest." Game is there—the skulking roebuck or even a scurrying *Rudel* of red-deer is to be seen any evening, even from a train-window. Streams and lakes are there, cleaner of cans and old tires than our own, and no worse beset with hotels and "bide-a-wee" cottages. But yet to the critical eye, there is something lacking that should not be lacking in a country which actually practices, in such abundant measure, all of the things we in America preach in the name of "conservation."[58]

The German "passion for unnecessary outdoor geometry" was Leopold's explanation for the "lack of wildness" in the landscape: "Most German forests, for example, though laid out over a hundred years ago, would do credit to any cubist. The trees are not only in rows and all of a kind, but often the various age-blocks are parallelograms, which only an early discovery of the ill-effects of wind saved from being rectangles." The border "between wood and field tends to be sharp, straight, and absolute, unbroken by those charming little indecisions in the form of draw, coulee, and stump-lot." Leopold was even more perturbed at what "the geometrical mind" had done to Germany's rivers: "If there were only room for them, it would be a splendid idea to collect all the highway engineers in the world, and also their intellectual kith and kin the Corps of Army Engineers, and settle them for life upon the perfect curves and tangents of some 'improved' German river. I am aware, of course, that there are weighty commercial reasons for the canalization of the larger rivers, but I also saw many a creek and rivulet laid out straight as a dead snake, and with masonry banks to boot."[59]

Germany's "cubist" forests were relics of the eighteenth century, when wood was indispensable for construction, shipbuilding, iron and glass manufacturing, and most other crafts and trades. With the bogey of "wood shortages" haunting Europe, Prussian bureaucrats and foresters (and their counterparts in Saxony, Hessen, and a few other "enlightened" German states) developed sophisticated methods of quantifying and maximizing the number of trees under their jurisdiction. Out of this impulse came the science of "mathematical forestry," the doctrine of "sustainable yields," the practice of monoculture—and the dawn of state-managed forests. Ultimately it also produced the even-bordered, even-lined, same-aged, same-species, growth-stunted, disease-ridden, ecologically impoverished "parallelograms" that Leopold saw in 1935. As Henry E. Lowood observed in an essay on eighteenth-century silviculture: "The German forest became an archetype for imposing on disorderly nature the neatly arranged constructs of science. Witness the forest [Johann Heinrich] Cotta chose as an example of his new science: over the decades, his plan transformed a ragged patchwork into a neat chessboard. Practical goals had encouraged mathematical utilitarianism, which seemed, in turn, to promote geometric perfection as the outward sign of the well-managed forest; in turn, the rationally ordered arrangement of trees offered new possibilities for controlling nature."[60]

Germany's "dead snakes"—to use Leopold's colorful description of the country's rectified rivers—were artifacts of the coal age, when the needs of mining, metallurgy, urban growth, and sewage disposal haunted industrial Europe. The Prussian government's role in managing its resources changed from the eighteenth to the nineteenth century: whereas a "cameralist" (state-sponsored) mentality prevailed in the wood age, a "liberal" approach was more in keeping with the coal age. Prussia privatized the Ruhr coalfields in 1865, and turned over its rivers and streams to dam and riparian cooperatives in the time span between 1891 and 1904. Institutions and actors also changed: whereas the foresters (and to a lesser extent the fisheries) were the arbiters of the state's vital resources in the preindustrial era, hydraulic engineers, urban planners, and industrial entrepreneurs took over those roles in the coal age. But the same utilitarian and geometric mentality was at work in both eras, despite the changeover in personnel. Once coal reigned supreme, the Prussian state began to view water in its rivers not as a habitat for fish but as a vital natural resource in its own right—a commodity that needed to be stored, pumped, and piped from reservoir to industry—just as it had earlier viewed its forests as so many planks of timber. Rivers were no longer a natural part of the landscape: they were just untamed canals or poorly designed drainage ditches, just as forests had once been inefficient producers of wood fiber. "The administrators' forest cannot be the naturalists' forest," Scott has pointed out in *Seeing Like a State*, because bureaucrats see no profit in overgrowth, foliage, grasses, mosses, shrubs, vines, flowers, and other features of a forest ecosystem. "Everything that interfered with the efficient production of the key commodity was implacably eliminated. Everything that seemed unrelated to efficient production was ignored."[61] Similarly, an industrialist's river is not an ecologist's river, for the industrialist sees only a water-commodity machine not a complex biological habitat. A water machine should be straight not curved, simple not complex, direct not circuitous, fast not slow—even if it is also dead rather than alive. A river is judged by the amount of water it delivers, not the number of fish species and aquatic plants it nourishes. With the triumph of the industrialist's vision, a typical German river came to resemble a stem-and-leaf mathematical diagram—the river equivalent of a cubist forest.

Prussia's supremacy over the Rhine and Ruhr (and after 1871 over the German Empire) helped to impose a unity of purpose "from above" on water management in the Rhineland and Westphalia. But the sole purpose was commercial and economic development, not protection of the river basins and riparian habitats. As the pace of industrial growth quickened, and as boats and barges expanded in size and number, the government responded by deepening the navigation routes, constructing new canals, and fortifying the Rhine's banks and harbors. But there was no corresponding effort to keep the river clean, protect the fish, or preserve the riparian vegetation. On these issues, the government opted for neglect. Similarly, the dam and riparian cooperatives owed their existence to special Prussian decrees. This legislation was needed in order to make membership in the associations compulsory, the only effective way of spreading the operating costs across all water users and water suppliers. But the same Prussian government, after coercing industries to join these cooperatives, took a laissez-faire attitude when it came to environmental protection. Far from placing a brake on resource exploitation, the government allowed these cooperatives to function like little despots, operating not democratically "from below" as their charters imply but from a position of strength that all but freed them of any restrictions.

Left to their own devices, Rhine-Ruhr industries arrogated for themselves the right to turn Westphalia and Rhineland's rivers into industrial faucets and sewage gutters. It was the cooperatives' directors, not the local citizens, who determined the shape and contours of the new riverscape; it was industries, not farmers, who got the lion's share of water and land; and it was engineers, not fishermen, who triumphed. "The conversion of the Emscher into a sewage network means, to be sure, that two-and-a-half million citizens have been deprived of clean-flowing brooks," explained August Heinrichsbauer, author of an early Ruhr hydraulic history and an avid apologist for coal and metallurgical firms, in 1936. "But this regulatory system was necessary in order to save the mining industry. All of the tributary beds had to be cut deeply into the earth in order to compensate for the ground sinkings, so deeply in fact that they were useful for nothing else. And in any case the water was laden with chlorides from the mines. Under these conditions it made no economic sense to build a second sewage system at an additional cost of 100 million Marks alongside an already existing natural one."[62] (As late as the 1950s schoolchildren in the Ruhr region were taught that Westphalia had performed a patriotic duty by sacrificing the Emscher to industry.)

Minimal cleansing at minimal cost: that was the cooperatives' motto and that was what the Prussian government was all too willing to tolerate. Invariably this meant devising massive dam-and-reservoir systems in the Rhine watershed to collect fresh water for use by industries and cities; constructing canal-and-river conduits to transport wastewater and other debris from industrial sites to the Rhine as efficiently as possible; and doing almost nothing to protect the Rhine itself as a source of clean water or as a riparian habitat. By 1945, the Rhine was digesting vast quantities of coal and lignite-related wastes from the Emscher, Lower Ruhr canal, Lower Lippe, Moers, Erft, Erft canal, Cologne rim canal, and dozens of smaller influx sites. The underlying assumption was that the Rhine's water volume was so vast, and its velocity so swift, that no amount of coal dust, phenol, chlorides, metallic tailings, or other carboniferous pollutants could possibly turn the river into a larger version of the Emscher and Erft.

Just how wrong that assumption was became apparent when a new economic power muscled its way onto the Rhine in the late nineteenth century: the chemical industry. "Mine: blast: dump: crush: extract: exhaust," was Mumford's curt summation of mining and metallurgy.[63] Burn, boil, vaporize, vent, spill, pollute, poison—these were to become the watchwords of the chemical age. The Rhine's coal and lignite fields lay primarily underneath Westphalia and Rhineland, so mining and processing were largely confined within a distinct and relatively small geographic boundary: the Lower Rhine. There were no such geographic limits on chemical production, especially once new energy sources—hydroelectricity, petroleum, and nuclear fission—became widely available in the twentieth century. These new sources allowed chemistry to colonize and pollute stretches of the Rhine that had once lain beyond the boundary of profitability.

There is no better barometer
to show the state of an industrial nation than
the figure representing the consumption of
sulphuric acid per head of population.

— Benjamin Disraeli (ca. 1860)

The chemist decides war and peace
with his discoveries.

— Otto von Bismarck (1894)

CHAPTER 5

Sacrificing a River

The worst industrial accident in Rhine history occurred at the Sandoz chemical plant in Basel-Schweizerhalle in 1986. It began when a storage facility burst into flames shortly after midnight on November 1, igniting over a thousand metric tons of insecticides, herbicides, fungicides, fertilizers, and other agrochemicals. Because the facility was not equipped with a modern sprinkler system, firefighters had to douse the blaze the old-fashioned way: by squirting hundreds of thousands of gallons of water onto the facility. The firehoses contained the blaze in a matter of hours, but the douse water flushed somewhere between ten and thirty metric tons of unburned chemicals into one of the plant's industrial sewers. From there the chemical brew found its way to the Rhine and began to scour the riverbed as it flowed downstream.

Among the released chemicals were the insecticides disulfoton, thiometon, etrimphos, and propetamphos, all highly toxic to fish. Within days the entire eel population from Basel to the Lorelei lay dead or dying. Grayling, pike, zander, and other fish species on the Upper Rhine survived only if they happened to be in a few sheltered backwaters and tributaries when the chemicals reached that stretch of the river. Even the fish that survived struggled in the aftermath, since the chemicals wiped out most of the microorganisms upon which they depended for nourishment. It took several weeks before urban sanitation facilities could begin drawing water from the Upper Rhine again, months before the river's microorganisms mounted a comeback, and years before the eel population returned to normal levels.[1]

Sandoz has found a niche alongside Minamata, Seveso, Love Canal, Bhopal, Chernobyl, and Exxon Valdez in the annals of environmental disasters. Yet for all its immediate destructiveness, the accident left no permanent mark on Rhine flora and fauna, no lasting impact on pollution levels, no indelible scars on the river's bank and bed. The Rhine's swift current flushed most of the contaminants into the North Sea within weeks. The full impact, moreover, was largely confined to the Upper and Middle Rhine stretches. Downstream sanitation networks in North Rhine–Westphalia and the Netherlands were shut down only briefly to allow the main chemical swell to pass through.

Far more harmful to the Rhine's long-term health has been the everyday discharge of industrial and urban effluents into its waters, the routine (and often illegal) dumping of chemicals into its channel bed, and the drip-by-drip influx of runoff from agriculture, roads, automobiles, and other human activities in its basin. These toxins, coming as they do on a continual basis, routinely kill off many of the microorganisms that keep the river clean, reducing the quality of the water to the point where only pollution-tolerant species can survive. Most riparian flora and fauna can recover from an occasional industrial accident: it is far more difficult for them to survive the everyday bombardment of poisons. The story behind the spill is therefore more illuminating than the spill itself: Why was the Sandoz chemical plant situated so close to the Rhine's bank? Why did its drainage canal lead directly to the Rhine channel? Why was a modern storage facility in safety-conscious Switzerland not equipped with fire sprinklers?

Europeans, of course, knew the hazards of common chemicals long before the industrial age. Oil of vitriol (sulfuric acid), aqua fortis (nitric acid), aqua regia (hydrochloric and nitric acids), lime, mordants, and tannic acid were all well-recognized toxins used in gold and silver smithing, leather tanning, cloth dyeing, and other commercial activities. Typically the manufacturers and users of these chemicals dumped their wastes directly into the closest streams without giving much thought to the downstream inhabitants or the organisms in the river, trusting in the well-recognized but little understood "self-cleansing" capacity of rivers. Neither environmental protection nor environmental cleanup belonged in their lexicons. But preindustrial firms also tended to be small operations, and the range of poisons they used was limited. The damage they inflicted on rivers was almost always confined to a small stretch that could be "sacrificed" to commercial activities with little danger or harm to the entire stream. Even large urban centers such as Paris could, in preindustrial days, resolve most of their river pollution problems simply by forcing the "riverside crafts"—tanning, dyeing, clothmaking, parchment-making, and butchering—to congregate a safe distance upstream or downstream from the city limits. Environmental protection in preindustrial Europe amounted to little more than the practice of sacrificing a small stretch of a river at the altar of commerce, and letting nature take care of the rest.[2]

Unfortunately, these slapdash practices continued into the nineteenth and twentieth centuries, at a time when there was a quantum leap in the production of known hazardous chemicals as well as new chemicals of unknown toxicity. Pollution-intensive industries, once confined to small river stretches, began to grow in size and proliferate in number so quickly that they sometimes encroached on entire rivers. By the 1870s, moreover, European water and hygiene experts—Alexander Müller, Louis Pasteur, Robert Koch, and others—had begun to realize at long last that the self-cleansing of rivers was accomplished mostly through bacterial processes, not through dilution, chemical interactions, and other mechanical processes as was previously assumed. By 1898, Carl Mez had devised a method for determining whether a river was "clean," "lightly polluted," "polluted," or "heavily polluted" based on a biological analysis of its water. Variations on Mez's method are still widely used today.[3]

The implications were unsettling for chemical manufacturers. If a river rids itself of pollutants biologically and not mechanically, then obviously an overdose of chemicals could negate its self-cleansing capability. Either manufacturers would have to alter their dumping practices or risk overwhelming a river system. Dumping, however, was free whereas pollution-abatement devices were costly to build and maintain. The chemical industry therefore found it more expedient to uphold the discredited solution-by-dilution method than to underwrite the costs of constructing purification plants. Company propagandists argued that *organic* pollutants from urban sewage networks, not *inorganic* pollutants from industry, were the main culprits in river pollution (some even made the outlandish claim that industrial effluents were beneficial to river water because they helped destroy organic pollutants). Most inventively, they adapted the old sacrificed stretches notion to their own purposes. "Sacrificed stretches," wrote Curt Weigelt, a biologist turned spokesperson for the German chemical industry, in 1901, "are spans of the river where pollution should be permitted either because the situation is such that industries can find no other possibility for getting rid of their wastewater without endangering profits and jobs, or because local conditions simply do not allow the cleansing of the water to its original state."[4]

Rivers, in other words, had to be sacrificed at the altar of progress, even at the risk of overwhelming their self-cleansing capacities. That the chemical industry would advance this line of reasoning, a mere variant of the rhet-oric of the coal and iron industries, was not surprising. That the German and Swiss governments took it seriously, and based their water policies on it, was symptomatic of a larger problem—the environmental shortsightedness that prevailed on the Rhine at the beginning of the twentieth century.

Military considerations also helped to reinforce the political myopia, especially in Germany, home to I. G. Farben (short for "Community of Interests of the Dye Industry"), a chemical cartel originally formed in 1925 to protect Germany's lead in world chemistry; it later became a silent part-ner in Nazi war crimes. Like coal and iron, chemicals gave enormous dip-lomatic and economic leverage to those states that produced them. At the outset of World War I, Germany's armament industry was dependent on trade with Chile for about half of its supply of nitrates (saltpeter), the chief ingredient in modern explosives. When Britain imposed a naval blockade on German ports, the German chemical industry turned to nitrogen-syn-thesis (using the newly developed Haber-Bosch process) and converted its dye-manufacturing plants into armament factories—without which Ger-many would have been defeated in months rather than years. Both poison gas, used in the trenches of World War I, and Zyklon-B, used in the gas chambers at Auschwitz and elsewhere, came out of the German chemical industry. So did synthetic fuel (benzine), synthetic rubber (buna), and arti-ficial fibers, all essential to Hitler's war-fighting capability. Regulating an industry that gave Germany an aura of potency and autonomy was not something German political leaders were inclined to do—at least not until after 1945.[5]

The Rise of Global Chemistry

Nineteenth-century industrial chemistry revolved around five basic prod-ucts: acids, alkalis (bases that dissolve easily in water), fertilizers, explosives, and dyes. All were produced on the Rhine.

Sulfuric acid was the only commercially valuable mineral acid in wide-spread use at the beginning of the industrial revolution (and it remains today one of the most widely sold chemicals on the world market). An oily and corrosive liquid, it was used in dye production and cloth bleaching, and it was needed by gold and silver smiths to derive hydrochloric and nitric acids. Until the eighteenth century, European manufacturers were largely dependent on a single source, the Saxon town of Nordhausen, for their supply of sulfuric acid. Though a hazardous chemical, its limited availability and exorbitant transportation cost acted as a powerful brake on the environmental damage it could cause. Then in 1736 a British entrepreneur succeeded in producing sulfuric acid in a laboratory by burning sulfur and saltpeter together. A decade later, another British entrepreneur built the first lead-coated chamber. These breakthroughs made it possible to manufacture sulfuric acid cheaply wherever sulfur and saltpeter could be found or hauled.[6]

Alkalis, chiefly sodium carbonate (or "soda ash"), were in high demand in Europe for soap production, textile dyeing, cloth bleaching, and glassmaking. Egyptian natron, Scottish potash, and Spanish barilla were the best-known mineral and vegetative sources of alkali, but they were expensive to process and transport. Demand therefore always far outstripped supply, keeping the environmental impact to a minimum. Then in 1791 the Frenchman Nicolas Leblanc invented a cheap process that used three common products—sulfuric acid, salt, and calcium carbonate (limestone)—to produce synthetic soda. Salt was first mixed with sulfuric acid to produce sodium sulfate; then the sodium sulfate was mixed with calcium carbonate to create soda ash as well as other commercially valuable products such as caustic soda and chlorine.[7] A century later, superior technologies replaced the Leblanc method. In 1873 the Belgian Ernest Solvay built the first ammonia-soda plant, which produced soda ash by the simple and cheap method of reacting ammonia with salt. Whereas in 1863 all 150,000 tons of Europe's sodium carbonate came from Leblanc factories, by 1902 over 1.6 million tons came from the Solvay method, accounting for about 90 percent of world production. Toward the end of the nineteenth century it also became possible to produce chlorine through electrolysis.[8]

British and French entrepreneurs all but monopolized research on acids and alkalis, as they did most chemical research, in the eighteenth century. But it was nineteenth-century German researchers who made the major breakthroughs in fertilizer and dye manufacturing, with direct implications for the Rhine. The three central figures were Justus von Liebig, whose pathbreaking research on soil nutrients for plants (ammonia, carbon, nitrogen, potash, soda, lime, sulfur, phosphorus) gave rise to the field of agrochemistry and to the production of synthetic fertilizers; Friedrich August Kekulé, whose discovery of the six-carbon benzene ring provided the theoretical backdrop for most coal-tar research; and August Wilhelm Hofmann, the first director of the Royal College of Chemistry in London, who identified the chemical properties of aniline (an alkaline oil from coal tar), a key ingredient in synthetic dye production.

The first practical breakthroughs in fertilizer production came with the discovery that phosphorus and nitrogen were the active ingredients in guano (bird excrement from the Peruvian coast), and that potassium was the active ingredient in wood ash (or "pot ash"). All subsequent research focused on producing ionic forms of phosphorus, nitrogen, and potassium for use as plant nutrients. The first phosphate fertilizers—known as super-phosphates—came into use in the 1840s. They were manufactured by combining sulfuric acid with phosphate rock or (after 1878) with "phosphatic slag" from the Gilchrist-Thomas steelmaking process. Potash (or potassium) fertilizer came into production after the discovery of salt deposits in Stassfurt (Saxony) in 1856. Output zoomed after the Germans took control of Alsace-Lorraine in 1871 and began to exploit the Alsatian potash mines to their full extent. Ideal for light sandy soil, potash was the fertilizer of choice for the sugar-beet industry. Nitrogen-based fertilizers first came into widespread use during the 1870s. This type of fertilizer was made from two ingredients—nitrates (mostly from Chile) and ammonium sulfate (from coal-tar distillation)—until the Haber-Bosch process made it feasible to synthesize ammonium nitrate in 1913.[9]

Nitrates were essential not only to fertilizer production but also to the modern explosives industry. Until the mid-1800s, gunpowder accounted for all blasting and firing devices. Then experimentation with nitroglycerine and nitric acid finally began to yield commercially usable results. Alfred Nobel's invention of dynamite (nitroglycerin and kieselguhr) in 1867 marked the turning point, and soon thereafter a whole range of nitrogen-based variants appeared, from ammonium-nitrate blasting agents to smokeless nitrocellulose propellants. Since these synthetic explosives lent themselves to everything from mining to tunneling to demolition and weaponry they found a ready market throughout Europe and the world.[10]

Synthetic dye production began in the 1850s with the discovery that coal tar, a byproduct of gasworks (city lighting) and metallurgical coking, produced a spectacular array of colors when fractionated and mixed with other chemicals. Before this discovery, dye firms were dependent on natural pigmentation derived from dyewoods, dye plants, insects, and minerals. Indigo, woad, cochineal, vermilion, brazilwood, and weld produced many of the common blue, red, and yellow dyes. But hues varied greatly between batches, as did colorfastness, and none of them worked well on the wonder fiber of the mechanized textile industry: cotton.

The first commercially viable synthetic dye was stumbled upon by William Henry Perkin, a Hofmann student in London, in 1859. It was an aniline purple, which he sold under the name "mauve." Soon thereafter competitor firms combined aniline with toluidine to produce the first aniline reds, patented as "fuchsin" in France and "magenta" in Britain. A wider palette of colors became available once Hofmann had completed his research on aniline: bleu de Lyon, Hofmann's violet, Bismarck brown, imperial purple, iodine green, among others. Other researchers discovered how to synthesize alizarin (the natural dye ingredient in madder root) from anthracene in 1869, and how to produce "azo" dyes from benzene and naphthalene in 1875. Soon a whole new variety of colors and hues with even greater colorfastness were available, including cachou de Laval, Biebrich scarlet, Congo red, and alizarin Bordeaux. Finally, in 1897, synthetic indigo (blue) came into production in Germany, after Adolf von Baeyer unlocked the structure of natural indigo in 1883.[11]

Acids, alkalis, fertilizers, explosives, and dyes formed the basis for nearly all nineteenth-century industrial chemisty. But research in these sectors also gave rise to a host of other new products, adding to the number of hazardous chemicals entering Europe's air and water. Most of the new products came via the dye industry, where chemicals were readily available, scientific expertise was concentrated, and investment funds easy to obtain. The popularity of synthetic dyes, for instance, created a demand for new coal-tar ingredients: benzene, toluene, xylene, naphthalene, phenol, cresol, catechol, hydroquinone, and naphthol. Before many of these ingredients could be utilized as dye intermediaries, they needed to undergo nitration and sulfonation processing, which in turn required ever-increasing quantities of nitric and sulfuric acid. Methods of synthesis were also applied to other dyemaking products, such as grain and wood alcohol, resulting in an array of new ingredients: formaldehyde, acetaldehyde, formic acid, acetic acid, ether, acetone, and alkyl halides. From there it was a short step to plastics, varnishes, perfumes, film, pharmaceuticals, synthetic rubber, synthetic gasoline, pesticides (chlorinated hydrocarbons and organic phosphates), and other products.[12] Chemical research also transformed paper manufacturing, sugar processing, and many other industries.

The Rhine as a Chemical Site

Many of Europe's most successful chemical firms settled on the Rhine or one of its tributaries, chiefly the Neckar, Main, and Wupper. The Rhine basin offered a steady stream of water, which was needed by chemical plants for production, for heating and cooling, and for dumping wastes. The river's superb navigational links, inland harbors, and seaports also played a major role. Sulfur, pyrite ore, salt, nitrates, chalk, limestone, phosphorus, bone meal, and other essential raw materials were either locally accessible or easily imported on Rhine barges. Meanwhile, Rhineland-Westphalia's coal provided a cheap fuel and a superabundance of coal tar, the chief ingredient in dye production. Acid, soda, and dye factories also found it profitable to settle in proximity to their best customers, the well-established textile mills on the Wupper, Ruhr, and Rhine. Superphosphate production relied on a ready supply of sulfuric acid as well as phosphate deposits in Rhineland and elsewhere nearby (such as Limburg on the Lahn tributary). Potash from Alsace, and phosphatic slag from Lorraine, also made the Rhine central to that sector of the fertilizer industry.[13]

A handful of Rhine firms can trace their lineage back to the early acid-and-soda days, but the takeoff period really came in the 1860s with the boom in synthetic dyes and (to a lesser extent) fertilizers. By 1875 there were 512 chemical plants in Rhineland alone, and hundreds of others elsewhere on the river.[14] Some firms consisted of one factory manufacturing a single product—sulfuric acid, ammonia, fertilizers, or dyes—but more typically a firm would consist of several interlocking plants manufacturing a wide variety of related products. Matthes & Weber (headquartered in Duisburg), for instance, began as a Leblanc-type soda factory in 1837, then added bleaching powder in 1840, ammonia in 1879, and sulfuric acid in 1905, never diverging much from its original soda-and-acid line. Weiler–ter Meer (Uerdingen), formed in 1893 from a merger of J. Weiler and E. ter Meer, continued the product lines—acids and dyes—of its former companies, while gradually adding new related products.[15]

Most of the early Rhine firms were under German and Swiss control (the Dutch would not establish a powerful presence until after 1945, when petrochemicals came into their own). The most successful of these firms grew into mammoth enterprises of worldwide renown. By 1914, for instance, German chemical production was largely in the hands of just four megafirms—BASF, Bayer, Hoechst, and Agfa—the first three of which were headquartered on the Rhine. BASF, founded in 1861, located its first plant in Mannheim, a right-bank Rhine city at the Neckar mouth. Following a dispute with the city over pollution issues, the company moved to the left-bank village of Ludwigshafen in 1865. The Hoechst firm (originally Meister, Lucius & Brüning) was founded in 1862 on the Main tributary near Frankfurt, where it has remained ever since. The Bayer company, founded in 1863, first settled on the Wupper in the textile town of Barmen, then moved downstream to the nearby town of Elberfeld in 1866 to avoid nuisance suits; it then moved its main branch to Leverkusen (near Cologne) on the Rhine in the 1890s.[16]

All three Rhine firms got their start in dye production in the 1860s. Proximity to the river gave them such a production edge that they quickly undermined their domestic competitors in Erfurt, Chemnitz, Leipzig, Stuttgart, and Nuremberg. (The Berlin-based Agfa, the only German giant not on the Rhine, survived by switching to photochemicals in 1887.) Just as the Rhine firms towered over the German chemical industry, so Germany also began to tower over the rest of the world. By 1913, German production of sulfuric acid stood at 1.6 million metric tons per year, while British production was 1.25 million tons. German dye production, meanwhile, reached a staggering 370,000 metric tons per year, representing over 85 percent of world market share—nearly twenty times higher than that of Britain.[17] Chemical production, especially dye manufacturing, was an enormous source of pride to the Germans, and helped foster the volatile idea that Germany was strong enough to challenge Britain's economic and political leadership worldwide.

Switzerland's earliest and most successful chemical firms—Ciba (Gesellschaft für Chemische Industrie Basel), Geigy, Sandoz, and Hoffmann–La Roche—were also situated on the Rhine, mostly hugging the Basel harbor, where they had access to raw goods, markets, and (after 1898) cheap hydroelectric power. Like the Germans, the Swiss began by imitating the French and British in the production of aniline dyes (usually fuchsin), relying on the technical know-how of their researchers to come up with innovations and other product lines. From the outset, however, the Swiss firms were at a competitive disadvantage, in part because they lacked a domestic supply of raw materials and coal-based chemicals, and in part because the Upper Rhine still posed insuperable navigational difficulties, a situation that would not change until the Kembs diversion dam was completed in the 1930s. Swiss firms managed to survive by allying themselves with the larger German firms, and by building a worldwide reputation in specialty dyes and pharmaceuticals, leaving the high-volume chemicals to the Germans. The largest, Ciba, founded in 1859, linked itself to BASF and specialized in certain aniline dyes and pharmaceuticals. Geigy (which later merged with Ciba), built an aniline plant in 1859, then soon thereafter began to concentrate on tanning extracts and wool dyes. Sandoz, founded in 1886, started in azo and sulfur-based dyes and then added pharmaeuticals to its product line. Hoffmann–La Roche, founded in 1894, specialized in pharmaceuticals from the outset.[18]

No other river in the world has been colonized to such a degree by chemical plants. The British chemical industry established an early presence in London and Lancashire, especially the coal-bearing region around St. Helens, northeast of Liverpool; it then spread to Tyneside, Glasgow, south Wales, and a handful of other sites. The Tyne, Mersey, Thames, and many other British rivers thus felt the impact of the industry, but not even the mammoth consortium Imperial Chemical Industries consolidated itself on a single river. France's chemical industry sprang up in three distinct parts of the country: the coal region of northern France; the coal-and-potash provinces of Alsace and Lorraine; and the textile district of Lyon, where the Saône flows into the Rhône. Only at Lyon did chemical firms (Rhône-Poulenc and Roussel-Uclaf) usurp a French river to a notable extent. The Russian and Soviet chemical industry started in St. Petersburg and Moscow-Gorky, then spread to the Donets-Dnieper region in the Ukraine; during and after World War II it spread eastward to Siberia, Kazakhstan, Central Asia, and the Volga region. Innumerable rivers were partially sacrificed in the process, including the Neva, Donets, Dnieper, and Volga. Chemistry in the United States has remained similarly decentralized. The industry concentrated at first in the Northeast, mostly between Maryland and Massachusetts, then gradually spread to all fifty states. Du Pont settled in Delaware, Dow in Michigan, Procter and Gamble in Ohio, Exxon (Standard Oil) in New Jersey and then Texas, Union Carbide in Connecticut, and Monsanto in Missouri. Many American rivers were thereby chemically burdened—not least the Upper Mississippi—but none even remotely as much as the Rhine.

The Environmental Impact

The Rhine thus left a permanent imprint on global chemistry. But industrial chemistry also left an indelible stain on the Rhine, for it was clear from the outset that the new factories were potentially hazardous to the river and those who lived on its banks. Chemical manufacturing typically begins with a basic feedstock chemical that is subjected to one or more reactions by adding other ingredients. In the nineteenth century, the feedstock was almost always a coal-based one, such as coal tar, but by the 1920s it increasingly became a petroleum-based one, such as olefin. Every chemical reaction potentially burdens the environment in several ways. First, in addition to the main product it nearly always creates one or more coproducts—often called "side products" if they are isomers of the principal product, "byproducts" if they can be used for other production purposes, and "waste products" if there are no known or profitable uses for them. Second, in nearly every reaction a residue of the basic feedstock remains at the end of the process; often this feedstock can be separated from the final product and reused, but sometimes not, in which case it is usually handled as a waste product. Finally, leaks, spills, and explosions caused by human negligence and mechanical failures allow products and coproducts to escape the production site.[19]

The first acid factories produced nitrogen oxide as a coproduct, which was discharged (along with the waste nitrogen in the lead chamber) into the atmosphere after production. The fumes were particularly harmful to the vegetation in close proximity to the factories, but they could also endanger the health of workers and neighbors, and they gave rise to a number of lawsuits. In 1827, Gay-Lussac developed a special condensation tower for nitrogen oxide, which worked by dripping small quantities of sulfuric acid to dissolve the nitrogen oxide gas before it went up the stacks. The tower, however, had one major economic drawback: it consumed sulfuric acid without producing any commercially valuable nitrogen coproducts, so firms had no incentive to invest in one except to settle nuisance litigation.[20]

In 1859, John Glover designed a second tower, which liberated the nitrous gas of its oxides of nitrogen and recovered the sulfuric acid for reuse. But the high installation cost outweighed the meager savings in reusable acid, so most companies were slow to adopt the new environmental measures, even when they were under legal obligation to do so. As late as 1883, for instance, German police discovered that the Carl Neuhaus firm (situated between Elberfeld and Sonnborn) had installed underground pipes that led from its sulfuric acid plant to the middle of the Wupper. Plant managers had instructed their workers to discharge waste chemicals into the Wupper under cover of darkness so that the acid would have a chance to reach the Rhine without being detected by the local authorities and residents.[21] It was not until the late nineteenth century that the combined Gay-Lussac and Glover towers came into general use in acid factories—nearly a half century after they had proved effective as environmental protection measures, and even then only after intensive government prodding.[22]

Noxious coproducts were also characteristic of the early soda (alkali) factories. The Leblanc method produced sodium carbonate by reacting common salt with sulfuric acid. Since one of the main coproducts, calcium sulfide, had no commercial value at the time, soda manufacturers let it pile up in the fields next to the factories. Or they dumped it in nearby streams where (as one British chemist observed in 1868) it was "carried away to the dumb fishes of the sea who cannot petition Local Boards or Parliament, nor bring a suit for nuisance."[23] Another coproduct, gaseous hydrochloric acid, had some commercial value as a liquid, but soda manufacturing produced it in such overabundance that it was handled as a waste and released into the atmosphere. It returned to earth in the form of acid precipitation (or "acid rain" as the British Alkali Inspector, Robert Angus Smith, first called it in 1872), killing vegetation and crops in the surrounding hillsides. In 1836, William Gossage developed a tower capable of capturing and recovering hydrochloric acid, but few industries rushed to adopt the method because supplies of hydrochloric acid far outpaced demand. Moreover, firms that did invest in a tower (usually to avoid lawsuits) continued to handle the recovered hydrochloric acid as a waste product and dump it into streams. Early use of the tower, therefore, only resulted in the preservation of hillside vegetation at the expense of riparian plant and animal life.[24]

Unlike other chemical sectors, alkali production became cleaner as the nineteenth century progressed (though the products themselves, chlorine and caustic soda, are anything but environmentally benign). In 1863, the Parliament of Great Britain passed the Alkali Acts, which forced soda factories to reduce their hydrochloric acid emissions by 95 percent. About the same time, new technologies made it commercially feasible to convert hydrochloric acid into chlorine for use as a bleaching powder in paper production. The high profitability of chlorine, in turn, ensured that the Gossage tower was adopted throughout Europe with or without regulatory measures. The Solvay method (which replaced the Leblanc method toward the end of the nineteenth century) was also more environmentally benign since sulfuric acid was not required in the manufacturing process. While not eliminating all waste products—calcium chloride, for instance, piled up on the outskirts of factories—it did reduce the overall burden of soda production on the environment. So too did the electrolytic plants, which produced chlorine and caustic soda with virtually no coproducts.[25]

Phosphate, potash, and nitrogen-based pollutants enter a river basin in two ways: from waste disposal at the production site, a problem that can be corrected by introducing better disposal methods; and from agricultural runoff, which can be corrected only by greatly reducing or eliminating the use of fertilizer near rivers. In superphosphate production, the main culprit is hydrofluoric acid, a gaseous coproduct that results from mixing phosphate with sulfuric acid. There were far fewer superphosphate plants than acid and alkali plants in nineteenth-century Europe, so hydrofluoric acid received less scientific attention and negative publicity than sulfuric and hydrochloric acid. But it was more toxic than hydrochloric acid and the damage it caused locally could be severe. One investigation revealed that a forest patch within an eight hundred meter range of a superphosphate factory had been all but destroyed by fumes: "Most of the pine trees circa fifteen years of age have died, while the birch are standing half alive among them. Some one hundred fifty meters northeast of the plant is another dying pine stand, and at four hundred meters the pine needles are brown." As with the acid and soda industries, a resolution of the problem came only after scrubbers were developed to capture the gas before it went up the smokestacks, and after local and national authorities forced companies to install these devices.[26]

Potassium and nitrate fertilizers pose different but related problems. Alsatian potash manufacturers (until forbidden to do so in the 1980s) released most of their waste chlorides (salts) into the Rhine, lending the water a slightly brackish quality harmful to many freshwater species. The manufacturing of nitrogen-based fertilizers (and explosives) creates several potentially obnoxious waste products: nitrates, nitrites, and ammonia compounds. If not disposed of properly, they leach through the soil and into groundwater and from there into lakes and streams, where they spur unnatural levels of algae growth.

Of all the branches of industrial chemistry, synthetic dyes had the most immediate and dramatic impact on air and water quality—and human health—in the Rhine basin. The early aniline dyes, for instance, required the use of arsenic acid for oxidation: the production of 100 kilograms of pure crystal fuchsin required 1000 kilograms of arsenic acid, 500 kilograms of "red oil" (aniline and toluene), 2100 kilograms of common salt, 200 kilograms of soda, 300 kilograms of nitric acid, 40 cubic meters of water, and small quantities of lime, caustic soda, and hydrochloric acid. Of the original 1000 kilograms of arsenic acid, approximately 140 kilograms disappeared in the dyeing process. About 260 kilograms of arsenic were recovered when the solid matter was mixed with cold water. The rest—600 kilograms, or

60 percent of the arsenic used in production—went out with the resin and wastewater.[27]

Waste arsenic was discarded without much thought to its impact on the environment or human health, at least until neighbors, and neighboring nations, began to complain. Friedrich Bayer, for instance, stored his arsenic waste in barrels next to his factory in Barmen. Eventually those barrels began to leak and burst, allowing arsenic to seep into the groundwater and poison the nearby wells. In 1864, after a major plant accident resulted in a serious arsenic spill, Bayer found himself on the losing end of innumerable lawsuits. Admitting defeat in 1866, he decided to build his new aniline factory downstream from Barmen, along a stretch of the Wupper which (in the words of a company biographer) was "already completely contaminated."[28]

Another dye manufacturer, Carl Jäger, also dumped arsenic directly into the Wupper until city residents and local authorities forced him to stop. Thwarted at home, Jäger began transporting his arsenic-laden barrels on Rhine barges and tossing them overboard into the North Sea. Between 1861 and 1863 alone, the firm dumped some 230 barrels containing 19,000 pounds of arsenic waste into the waters between Rotterdam and Liverpool. This absurd practice came to a halt in 1868, after the Netherlands and Prussia stepped in and signed a special international treaty that restricted the transport of hazardous chemicals on the Rhine and outlawed the dumping of arsenic in the North Sea.[29] These restrictions, however, did not apply to dumping in the Rhine channel itself. So when Jäger applied for a license to build a new factory in Lohausen (now part of Düsseldorf), the Rhineland government agreed despite the firm's miserable environmental record, stipulating only (as a sop to public opinion) that Jäger dispose of his arsenic via watertight pipes leading to the Rhine.[30]

The same kinds of shortsighted disposal practices prevailed in Switzerland. J. J. Müller-Pack, a Basel firm founded in 1860, poured its arsenic-laced wastewater into a nearby canal, where it seeped into the groundwater and wells, poisoning many Basel residents as the arsenic inched toward the Rhine. The local authorities—conscious of the health problem but not wanting to thwart a profitable new industry—decided that the best solution was to "export" the arsenic downstream, where it would become France's and Germany's problem. They therefore coerced all Basel-based chemical plants either to relocate their factories directly on the banks of the Rhine or to construct pipelines from their factories to the river. Since Müller-Pack could not afford the costs of pipeline construction, the firm began transporting its arsenic waste to the Rhine in barrels and dumping it from the middle of the Basel bridge. The larger firms—Ciba, Sandoz, and Geigy (which bought the Müller-Pack firm in 1864)—could afford more discreet ways of dumping their chemical wastes, but for the most part they all followed the same practice: pouring untreated arsenic compounds into the Rhine just downstream from Basel's city limits.[31]

The arsenic poisonings continued until Hoechst researchers discovered an arsenic substitute, nitrobenzene, in 1872, after which government officials in Basel and other cities coerced their dye manufacturers to switch production methods.[32] Unfortunately, arsenic was not the only toxic substance used in dye production. Hydrogen sulfide, sulfitic acid, arsine, aniline vapors, and sulfuric acid all poisoned the air inside and outside the factories, causing workers and locals to become sick. Coal-tar derivatives—phenol, xylene, toluene, nitrotoluene, benzene, and even nitrobenzene (the much touted arsenic substitute!)—were also dangerous to one degree or another. When absorbed by the human body in small doses over time, these chemicals could damage the liver, kidneys, and bone marrow, a fact that dye workers came to learn firsthand.[33] "The dye impregnates the skin of the people working in the alizarin block at Leverkusen, and at night some of the pigment gets sweated out through the pores onto sheets and clothing and then into the Sunday wash, after which it is impossible to dislodge the famous coloring," wrote Egon Edwin Kisch, in his exposé of the giant chemical cartel, I. G. Farben, in 1927:

> Take an evening stroll in Leverkusen and you will shake blue hands, green hands, red hands. It is most unfortunate that chemicals designed to impregnate textiles actually demonstrate their durability on the skins of living humans. Worse yet are the dye vapors in the lungs, the risk of lead poisoning, the mental illness caused by arsenical hydrogen gases in the lithopone unit, the acid burns and eye wounds in the place where zinc sludge is mixed, and the chlorine which eats away at the brain cells.... The only thing a spy needs to ascertain the groundplan of Leverkusen is a nose: the reek of sulfur, ammonium, oleum, acetylen, natron vapors, and chlorine gas, not to mention pesticides such as Uspulum, Solbar, and Certan (designed to kill insects but perhaps improvable for use against humans too), will tell him what lies where behind the factory fence.[34]

The dyes, of course, penetrated not just into the skin and nostrils of workers but also into surrounding rivers and streams. By 1902 the BASF plant in Ludwigshafen had six pipes that led to the Rhine, emitting enough red dye that a distinct trail was visible all the way to Worms several kilometers downstream. By 1905 Bayer had ninety-five pipes dumping wastewater into the Wupper, one of the many reasons it earned the nickname "river of ink." The Wupper "is so dark with filth," quipped Social Democratic Party leader Phillip Scheidemann in 1904, that a National Liberal who falls into it will "resurface as black as a Catholic Centrist." The Hoechst plant created similar problems on another Rhine tributary, the Main. It was producing so many different dye colors, said Scheidemann, that a dip in the Main river "will turn a white bathing suit rainbow-colored."[35]

Other Chemical-Based Industries

Adding to the chemicals being dumped into the Rhine and its tributaries was the newly emerging pulp-and-paper industry. Paper products are made from water and plant fibers such as flax, hemp, cotton, straw, jute, esparto, and wood pulp. Cellulose, the key ingredient, is extracted by separating the protoplasm from the fibers and then mixing it with water and rolling the mixture into sheets.[36] Papermaking originated in China two thousand years ago, but modern pulp-and-paper manufacturing owes its production techniques to nineteenth-century European chemists. A first step came in 1854 when a French researcher discovered that a chemical pulp could be obtained by heating straw and caustic soda. Shortly thereafter an American firm began manufacturing soda pulp by boiling wood chips in an alkali solution. Within a few years another American firm patented the first method for producing bisulfite pulp, and in 1879 a German chemist developed a sulfate pulp process that produced a highly durable paper from both soft and hard woods. Regardless of method, the resulting cellulose had to be treated further with chlorine or bleaching powder to remove the remaining impurities and to attain whiteness.[37] Boiling, washing, pulping, and bleaching all created persistent and pernicious water problems around the mills. This was especially true of bleaching waste, which if not handled properly could wipe out a river's entire fish population overnight.[38]

Easy access to raw fibers, acids, alkalis, and chlorine made Rhine tributaries prime locations for pulp-and-paper mills, especially once the wood pulp industry (which required large quantities of coal, wood, and water) came into its own in the 1880s.[39] Two minor Rhine feeder streams—the tiny Strunde and Neffel—were among the first to feel the impact. The Strunde was a crystal-clear mountain stream until the J. W. Zanders company (headquartered in Bergisch Gladbach) built a paper mill on its banks in 1829. By the mid-1830s the river's fish population was already endangered, and by the 1850s downstream communes were complaining of bluegreen foam and slime floating atop the stream. Modest efforts to restore the Strunde came to naught, in part because Zanders kept building more and more pulp-and-paper mills, and in part because wool spinning and lignite mining also gained a small foothold on the Strunde. For a time there was talk of constructing an industrial canal to the Rhine, but it proved too expensive to build, so the Strunde remained the primary dumping site for the region. "The stream's water appears light, pure, and clear until it reaches the first paper mill," noted a Cologne official in 1879. "After that the water gets murky, black and leaden in color, and on occasion a disgusting yellow foam piles up three to four feet high at the mills. Gas bubbles accumulate on the banks, or more correctly, on house walls near the banks, and they give off a horrid stench, especially when sunlight hits them, forcing the people who live in those houses to keep their windows shut." He added: "the water's dark coloration continues almost to the point where it enters the Rhine."⁴⁰ The Neffel was similarly overburdened with pulp-and-paper waste. At one point, a prominent resident of the region, Freiherr von Geyr zu Berg Müddersheim, won a lawsuit against the river's greatest polluter, the Zülpich Paper company. But the court imposed sanctions so weak that they did not deter Zülpich or any other paper firm from continuing to use the stream as a dumping site, and so by 1913 the Neffel—like the Strunde before it—had acquired a disgusting odor and lost its fish populations.⁴¹

The sugar-beet industry—another offshoot of modern chemistry— caused further environmental problems on Rhine tributaries. Cologne was a minor center of cane sugar refining before the nineteenth century, mostly because the Dutch dominated the world's sugar cane trade and some of that product reached Cologne via Rotterdam. In 1747 a member of the Berlin Academy of Sciences discovered that beet roots had a high sugar content, and by the early nineteenth century the first beet-sugar factories were built on German soil, mostly in Silesia, Saxony, and Thuringia. Beginning in the 1850s, factories also began to spring up in Rhineland, Westphalia, and Hesse-Nassau. While the Rhine firms never seriously competed with other German producers (and certainly not on the global market), they nonetheless collectively brought nearly 46,000 hectares under cultivation by 1938, about 9 percent of Germany's total beet production. Much of this land was former Rhine floodplain enriched by potash fertilizer.[42]

But more than floodplain loss and fertilizer runoff was at stake. Until recent technological improvements were put in place, beet-sugar manufacturing required between 7 and 27 cubic meters of water per ton of beets, the variation depending on how any given factory washed, sliced, diffused (separated the juice from pulp), saturated, sulfurized, evaporated, concentrated, and cured its beets, and whether or not it recycled its water. Since beet crops were harvested in the autumn, most processing took place between October and December, when Europe's non-Alpine rivers typically run low. This meant not only that the beet-sugar industry had to compete with other industries for river water but also that its wastewater was dumped back into the stream at a time when the flow was sluggish. Even under the best of conditions untreated sugar waste spawns bacteria and fungi, depleting the oxygen content of the water and producing hydrogen sulfide, methane, and other foul-smelling gases. Oxygen depletion (eutrophication), in turn, endangers fish populations by hindering their breathing; even when it does not destroy them directly, it kills off the nutrients upon which they depend. For decades the beet-sugar industry at Euskirchen generated a "fungus trail" 30 kilometers long on the Erft—and this on a river already choking with pollutants from lignite mining.[43] Other Rhine tributaries had similar problems with fungus trails and eutrophication as well.

The Rhine as a "Sacrificed Stretch"

The growth of the chemical industry did not go completely uncontested. Arsenic deaths, factory leaks, accidental fires, intermittent explosions—these all made the front pages of newspapers and helped to generate protests and lawsuits. Every new factory or expansion plan aroused a sense of unease among nearby residents, and frequently one or more groups tried to stop licensing and construction. In Rhineland alone, there were hundreds of such efforts between 1860 and 1914 (the heyday of chemical growth there), with the most intense fights occurring in densely populated regions—Barmen, Elberfeld, Cologne, and Düsseldorf. Ultimately, however, the protesters found themselves fighting rear-guard actions against the ever-quickening march of chemical "progress." Rare were the times when Prussian state authorities chose not to license a chemical plant, and even rarer were the times they shut down a factory for violating nuisance laws.[44]

Faced with mounting public concern over poisonings and other accidents, the chemical industry began to close ranks and promote the "sacrificed stretches" theory. In Germany, the Association for the Protection of the Interests of the Chemical Industry (Verein zur Wahrung der Interessen der Chemischen Industrie), and in Switzerland, the Society of Chemical Industry in Basel (Gesellschaft für Chemische Industrie in Basel), provided the organizational framework for cooperation. The two leading spokespersons were both Germans: Carl Duisberg, head of the Bayer Chemical Company, and Curt Weigelt, chair of the German Association's Commission on Wastewater. Weigelt also chaired Bayer's Special Wastewater Commission, which was charged with the task of "collecting material on the firm's wastewater circumstances and its influence on the Rhine." [45]

Industrial chemists were particularly avid champions of the "sacrificed stretches" idea because it dovetailed so well with their political and economic interests. If large rivers were capable of recovering from industrial pollution by themselves, then federal intervention to protect the Rhine was not merely superfluous but downright harmful to economic growth. Local problems, they insisted, should be handled on a river-by-river basis through riparian cooperatives like those already established in the coalfields of Rhineland and Westphalia, not via national legislation or an international agreement.

Matters came to a head when the Prussian parliament began debating the Prussian Water Act of 1913. The main purpose of this legislation was to subsume the seventy-odd separate decrees that governed Prussian water policy (some of which dated back to the eighteenth century) into one omnibus bill. A secondary purpose was to amend those laws to favor industrial development over agriculture. [46] But with so many Westphalian and Rhineland rivers choking with industrial effluents, and with so much discontent among citizens about the odors and poisons, many Prussian legislators were leaning toward more stringent wastewater regulations—a prospect that alarmed the chemical industry. "May God protect us from federal legislation, especially from a national wastewater law," Carl Duisberg flatly declared in 1912: "Water makes or breaks the chemical industry. A single plant on the Upper Rhine utilizes two hundred thousand cubic meters of water per day." Concerns over the Rhine's health were greatly exaggerated: "You could pour Germany's entire production of sulfuric acid into the Rhine at Cologne for an entire year and you would not find a trace of it downstream at Mülheim. The Rhine is capable of neutralizing seven million kilograms of sulfuric acid per day." A Berlin-based chemical firm, he claimed, had planned to build a branch on the Lower Rhine, but then decided against it when Prussian regulators "demanded that its wastewater be clean enough to keep trout alive.... It is simply absurd to expect fish to be able to survive in wastewater." The firm moved instead to Baden, "where it is child's play" to get a license. "This kind of competition among states is a good thing," Duisberg concluded, suggesting that other firms would also move upstream if Prussian authorities dared to impose stringent regulations.[47]

The Prussian government took the hint. The only significant initiative to come out of the wastewater debate was the establishment of a Clean Wupper Cooperative (Verband zur Reinhaltung der Wupper) in 1913. Here the Prussian authorities felt compelled to act after a special commission determined that the entire Wupper had become so polluted that "its inflow into the Rhine could be distinguished by its dark coloration, particles, decomposing organisms, and foamy water."[48] The Wupper Cooperative was similar in scope and purpose to its Emscher and Ruhr counterparts, except that textile and chemical firms rather than coal mines and steel mills dominated. As elsewhere, the state's guiding hand all but disappeared as soon as the legal machinery was in place, leaving local industries and communes to build the necessary dams and sewage plants as they saw fit.

Self-management, in this case, meant that it would take another seventeen years—until 1930—for the cooperative to function effectively. It also meant that Barmen and Elberfeld, the two largest upstream textile towns, were free to conspire with Bayer, the leading chemical company, to ensure that minimal cleansing at minimal cost was the top priority ("We can only gladly welcome these colors, which the factories discharge into the river," the mayor of Elberfeld once proudly proclaimed, "for only as long as the Wupper is dirty, is there still work to be found").[49] The Wupper Cooperative, therefore, was no more interested than its counterparts on the Ruhr and Emscher in restoring the river as a riparian habitat; all it was really designed to do was to apportion Wupper water to its members and maintain a level of water quality commensurate with industrial production. "Anyone familiar with the Wupper," the *Solinger Tageblatt* commented dryly, "knows that Elberfeld, Barmen, the Bayer factory, and a few other firms utilize the river to get rid of their waste products cheaply.... In truth they do not want a 'clean river cooperative.' They want a 'polluter cooperative' in which the financial burden for their measures (and halfway measures at that!) will be borne by the victims downstream." [50]

The *Solinger Tageblatt* was all too correct about the distribution of costs, but it was largely wrong regarding the distribution of benefits. For all its limitations, a cooperative at least had the advantage of bringing all water suppliers and water users under one administrative roof, where common issues could be discussed and bargained collectively. Even the most despotic cooperative was better than no cooperative: the alternative, after all, was not a natural and free-flowing stream but unchecked growth and exploitation. The Sieg, Bröl, Donner, Mühlen, Sülz, Frechen, Bachem, and Faul—all Rhine feeder streams in Westphalia and Rhineland—were polluted by the early twentieth century. All were investigated by Prussian water experts, yet all were deemed too small and inconsequential to warrant government action. That only meant, however, that each of these streams was deemed to be an acceptable sacrifice and that industrial-urban pollutants would be allowed to reach the trunk system—the Rhine—with virtually no environmental safeguards. [51]

The Sieg was the most extreme case among the unprotected rivers. Relatively small, less than 150 kilometers long, it flows through Rhineland (now serving as a border between North Rhine–Westphalia and Rhineland-Palatinate) before merging into the Rhine near Bonn-Beuel. Never central to the coal and metallurgy industries like the Emscher and Erft, or to chemicals and textiles like the Wupper, it was deemed too inconsequential to be worthy of a riparian cooperative. Nonetheless, a wide variety of industries managed to gain a toehold almost everywhere on its banks— iron ore factories between Weidenau and Wissen, textile mills in Eitdorf, and chemical, dye, metallurgical, and metalworking plants in Siegburg and Troisdorf. A slow-moving stream, the Sieg absorbed the effluents from these factories and carried them to the Rhine as best it could, but after a few decades of abuse its self-cleansing capacities were strained to the limit. In 1899, local residents started complaining about the "red-colored water" emanating from a Siegburg textile factory. By 1912 there was widespread concern about a plate-rolling mill in Wissen, by 1938 about a spun-rayon plant in Siegburg, and by 1947 about a chipboard factory in Etzbach. These problems, however, stayed below the radar screen of the Prussian state and so nothing was done to regulate industrial growth there, and as a consequence (to use Johann Paul's apt phrase) the "sacrificed stretches kept getting longer." Only a handful of "sacrificed stretches" existed in the 1890s.

By 1912, they had multiplied to such an extent that they covered half the river. And by the 1950s the entire river had been sacrificed to industry—all without benefit of a riparian cooperative.[52]

The Spread of the Chemical Industry

Coal was the driving force behind the Rhine chemical industry for most of the nineteenth century. It was coal that fueled the steam plants that powered the chemical factories, just as it was coal that fired the steamships that transported raw materials and finished goods on the Rhine. It was also coal byproducts that provided the basic feedstocks for nearly all chemical production. This dependency meant that almost all of the early Rhine chemical firms clustered on the Lower and Middle Rhine, and on the Neckar, Main, and Wupper tributaries, where coal was plentiful and the transportaton network thickest. By the twentieth century, however, new forms of energy were coming into their own. Hydroelectric power, petroleum-based fuels, and nuclear fission allowed the chemical industry to invade river spaces that had largely been spared the ecological damage of industrialization.

Mumford, a vociferous critic of the "paleotechnology" of coal and iron, became a major booster of the "neotechnology" of electricity and petroleum. Nearly everything that smacked of innovation enamored him: new metal elements such as aluminum and tungsten, new chemical elements such as radium and uranium, new synthetic materials such as asbestos and artificial silk, new forms of locomotion such as the gas-driven engine and airplane. Above all, he embraced the "clean" energy of hydroelectricity: "The smoke pall of paleotechnic industry begins to lift: with electricity the clear sky and the clean water of the eotechnic [windmill and waterwheel] phase come back again: the water that runs through the immaculate disks of the turbine, unlike the water filled with the washings of the coal seams or the refuse of the old chemical factories, is just as pure when it emerges."[53]

His optimism proved only partly justified. While hydroelectricity generates electricity without pollution, it causes many disturbances to river ecology. The building of dams entails the flooding of forests and meadows, the destruction of riparian habitat, the transformation of vegetation and animal types, and in some cases the submersion of entire towns and villages. In-river dams are particularly lethal to migratory fish—most famously salmon—because they sever the link between a river's mouth and its headwaters. Dams also trap agrochemicals, pesticides, heavy metals, and other toxic substances, turning reservoirs into chemical time bombs. Petroleum, meanwhile, burdens the atmosphere much like coal and most other fossil fuels: it just replaces the visible soot and smoke with the invisible pollutants of the gas engine. Similarly, nuclear energy (not yet on the horizon when Mumford wrote) brings with it the king-sized headaches of thermal pollution, waste disposal, and radiation risks. Not only are these energy sources less "clean" than Mumford imagined, but they also generate more industrial and urban growth, and thus more pollution and congestion problems, wherever they are utilized.

Hydroelectric development on the "Swiss Rhine" (Alpenrhein, Aare, and High Rhine) commenced in the 1890s, almost as soon as the technologies for constructing large dams and long-distance transmission lines had been developed. Emil Rathenau, head of German General Electric (Allgemeine Elektrizitäts-Gesellschaft), was the first to grasp the High Rhine's potential. Noting that "energy which elsewhere has to be wrenched from the earth is here for everyone to see," he supervised construction of a large dam

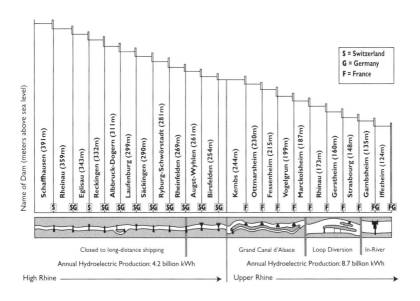

Figure 5.1 Hydroelectric production on the High and Upper Rhine today (Source: Garritsen, Vonk, and de Vries, eds., *Visions for the Rhine*, 21)

at Rheinfelden in 1898. Since there was at the time no industry on the High Rhine to consume the electricity, he stimulated demand by cofounding the Aluminium-Industrie Neuhausen (now Aluminiumhütte Rheinfelden) and convincing Elektrochemische Werke Bitterfeld (later Dynamit Nobel Werk Rheinfelden) to build a branch at the new dam site. Other chemical and electrochemical firms soon followed suit—Elektrochemische Fabrik Natrium (later Degussa Werk Rheinfelden), Hoffmann–La Roche, and Geigy—and within a decade the once sleepy village of Rheinfelden had become a thriving industrial center.[54] Rheinfelden, in turn, spawned ten other High Rhine hydroelectric dams between 1912 and 1963, each harnessed in one way or another to chemical, electrochemical, and aluminum industries.[55]

Protests from Swiss and German preservationist groups briefly threatened to slow the construction of one of these dams—at Laufenburg—site of the Rhine's most spectacular salmon runs, in the early twentieth century. But nature protection was still in its institutional infancy, and its leaders were easily written off as backward-looking Romantics. After a public debate, Baden's minister of the interior concocted a "compromise": he commissioned Gustav Schönleber of the Karlsruhe Art Academy to immortalize Laufenburg's "incomparably beautiful landscape" in a painting and then promptly gave the go-ahead for the dam's construction. Schönleber's painting still hangs in the Staatliche Kunsthalle in Karlsruhe—its haunting image of vanished beauty serving as a reminder of the hubris that governed Rhine affairs a century ago.[56]

Hydroelectric development also occurred on a more modest scale on the Alpenrhein tributary system in the Swiss canton of Grisons and the Austrian province of Voralberg. Since the Alpenrhein lacked the necessary steepness for the production of hydroelectricity, small dams were placed on its main feeder streams, the Vorderrhein, Hinterrhein, Tamina, and Ill. The first three—built by Elektrizitätswerke Bündner Oberland AG at Waltensburg (1908), Küblis (1919), and Klosters (1922)—fared poorly in the early years, due to the absence of urban and industrial customers. But the utility company's fortunes slowly reversed as cheap electricity drew more and more chemical-related industries to the region (not to mention tourists drawn to the new electric ski lifts), and by the 1930s an energy deficit had begun to develop. To fill the gap, the consortium Kraftwerke Hinterrhein (Thusis) undertook a series of new power stations at Ferrera, Bärenburg, and Sils in the 1940s; and Kraftwerke Vorderrhein AG (Disentis) started construction on its Sedrun and Tavanasa plants in the 1950s. The region's single biggest power consumer, and industrial polluter, was Emser Werke AG, a chemical plant located in Reichenau near the point of merger between the Vorderrhein and Hinterrhein. Modest in size compared with the plants on the Middle and Lower Rhine, the Emser factory specialized first in wood saccharification (an energy-intensive process that produces alcohol) and then branched out to include a wide variety of chemicals, fertilizers, and synthetic materials.[57] Downstream from Reichenau, construction of the Kraftwerke Sarganserland and Voralberger Illkraftwerke helped turn the towns of Bludenz, Feldkirch, Dornbirn, and Bregenz into minor centers of chemical, electrical, machine-building, pulp-and-paper, leather, and textile production. Much of the region's hydroelectric production eventually came into the hands of the consortium Nordostschweizerische Kraftwerke, now Switzerland's leading producer of Rhine-based electricity.[58]

The Treaty of Versailles (1919), meanwhile, provided the French government with the legal fig leaf it needed to construct a series of hydroelectric diversion dams on another stretch of the river, the Upper Rhine. With these new energy sources came new opportunities for industrial development in Basel, Baden, and (above all) in the newly reacquired Alsace-Lorraine. Strasbourg, Mulhouse, and Colmar were already centers of Alsatian agriculture, textiles, and mining before the hydrodams were built, but it was only after L'Energie Electrique du Rhin (now part of Electricité de France) constructed its first diversion plant at Kembs in 1932 that industrial sprawl began to hit the Upper Rhine in earnest, with eight more hydrodams coming on line in subsequent decades. Aluminum, automobile, cellulose, and a variety of other energy-intensive industries were enticed to Alsace to take advantage of the cheap hydropower and—once the Grand Canal d'Alsace was completed—the Rhine's excellent transportation network (see figure 5.2).[59]

By the 1920s it was already becoming clear to many industrial leaders that the Rhine's hydroelectric potential would be tapped out long before the Rhine's ever-growing energy needs were met. So they began to look to petroleum as an alternative energy source. As with hydroelectricity, the chemical industry provided much of the impetus, though the rapid growth of the automobile industry proved equally pivotal. The major breakthrough came when olefin, a coproduct of gasoline refining, began to replace coal as the basic chemical feedstock in the United States in the 1920s. European petroleum firms built a few Rhine refineries in the 1920s, but the first true

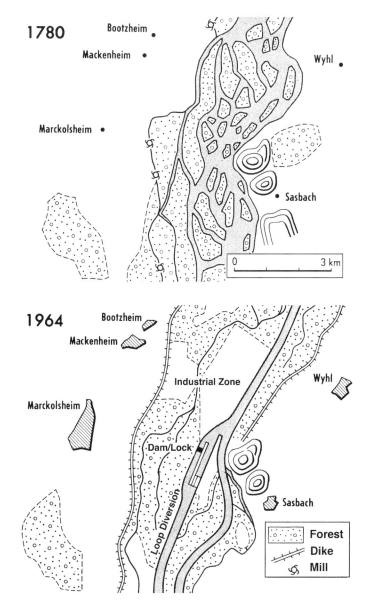

Figure 5.2 The Upper Rhine at Marckolsheim, in 1780 and 1964. Hydroelectricity helped transform Marckolsheim and many other small Rhine towns into modest-sized industrial centers. (Adapted from Juillard, *L'Europe Rhénane*, 133)

petrochemical factory—the Olefinwerke at Wesseling (near Cologne), a joint venture between BASF and Shell—did not come on line until 1953. Bayer and British Petroleum built a similar plant in Dormagen in 1958, while Hoechst procured its olefin supplies from new refineries in Kelsterbach and in Cologne.[60]

Ultimately, the Netherlands was the main beneficiary of the petroleum age, and that meant that yet another stretch of the river—the Delta Rhine—was sacrificed to the chemical industry. In the early 1950s, Rotterdam's port authorities began widening and deepening the Rhine mouth to accommodate Mideast oil tankers. They also constructed an oil harbor, "Europoort," to provide space for the Shell, Exxon, Nerefco (BP/Texaco), Kuwait Petroleum, and Eurosplitter/Eurostill refineries—sacrificing in the process two Dutch villages and a nature conservation park. Soon thereafter the Rotterdam-Rhine and Rotterdam-Antwerp pipelines were built so that crude oil could be piped rather than shipped upstream to German and Belgian refineries. Most important, recognizing that the refining and transshipment of oil was not a major revenue producer, Dutch authorities lured a total of thirty major chemical firms to Rotterdam and its surrounding cities, thereby helping to create "Randstad Holland," an urban conglomerate of eleven million people and one of the most densely populated regions in the world. Today, the Nieuwe Waterweg at the Rhine mouth—especially the 40-kilometer stretch between the old Waalhaven and the newly reclaimed Maasvlakte—is a seamless web of oil tankers, petroleum refineries, petrochemical plants, storage facilities, and port terminals.[61]

By the 1960s it was clear to many industrialists that even petroleum was not keeping pace with the Rhine's energy needs, and the oil crisis of 1973 further underscored Europe's fragile dependence on the volatile politics of the Middle East. Increasingly the riparian governments turned to nuclear energy for their salvation, and by the mid-1980s the Rhine was on its way to becoming one of the world's most nuclearized rivers. As a sign of the times, in 1966 the Swiss abruptly abandoned a half-built hydrodam at Koblenz (Switzerland) on the High Rhine and constructed instead four nuclear reactors near the Aare-Rhine merger point: Beznau I and II, Gösgen-Däniken, and Leibstadt, the last of which came on line in 1984.[62] France, meanwhile, took the lead in the construction of the Fessenheim I and II reactors on the Upper Rhine's left bank in Alsace, in cooperation with Switzerland and Baden-Württemberg. France also built four reactors in Lorraine, at Cattenom on the Mosel. The German nuclear industry, which languished at first due to world fears of a revived militarism, came into its own after West Germany joined NATO in 1955. The first commercial reactor was built on the Main in 1961 at Kahl. Then came Philippsburg I and II and Biblis A and B, all on the right bank of the Upper Rhine, and Neckar-Westheim I and II on the Neckar tributary. The Germans also took the lead in the construction of a breeder reactor at Kalkar on the Lower Rhine, built with the assistance of Dutch and Belgians. The Dutch, meanwhile, built a nuclear reactor at Dodewaard on the Delta Rhine.[63]

Europe's largest energy and chemical companies were the principal backers of the Rhine nuclear reactors, with Electricité de France, RWE, and Hoechst in the forefront. Their goal was to turn the Upper Rhine into a "second Ruhr" by exploiting nuclear rather than coal resources. Widespread public concern about the danger of radioactive contamination near the plants, however, soon spawned a number of protest movements, some of which grew in strength to the point where they could seriously challenge these megacorporations. In 1970, Jean-Jacques Rettig established France's first antinuclear group, the Comité de Sauvegarde de Fessenheim et de la Pleine du Rhin. The group led an unsuccessful attempt to block construction of the Fessenheim reactor in Alsace. (An attempt in 1975 by an unknown terrorist organization to blow up the plant also failed.) In 1972, Helga Vohwinkel initiated a legal battle in Germany to halt construction of the Mülheim-Kärlich reactor in Rhineland-Palatinate (a post-1945 state composed of the Palatinate and the southern half of Rhineland), after she uncovered a report that raised serious safety-related questions. She won a postponement in 1977 when a Rhineland-Palatinate administrative court concluded that there had been collusion between plant promoters and licensing authorities. Mülheim-Kärlich briefly went into operation in 1988, but was shut down within a year (this time permanently) because of seismic-related safety concerns.[64]

Meanwhile, various protest groups in France, Germany, and Switzerland formed an umbrella organization—the Rhine Valley Action Committee—in the early 1970s to coordinate their fight against the Upper Rhine's nuclearization. The Action Committee's first success came when it thwarted construction of a German-owned chemical plant that was planned for the French side of the Upper Rhine at Marckolsheim (the German company had set its sights on Alsace after failing to get a license at home). Its second success came when it thwarted construction of a nuclear reactor at Breisach in Baden-Württemberg by mobilizing farmers and villagers under the banner "Better active today than radioactive tomorrow!" Its third success came after the Baden-Württemberg government decided to build a nuclear reactor at Wyhl, near Breisach and Freiburg. This time there were frequent violent confrontations between demonstrators and police before a state administrative court finally quashed the reactor project in 1977.[65] After Wyhl, the Rhine chemical and utilities giants found themselves in a downward slide they could not stop. The jointly sponsored German-Dutch-Belgian breeder reactor at Kalkar never went into operation because of massive protests there. For similar reasons, nuclear plants planned for Schwörstadt, Kaiseraugst, Gambsheim, Wörth, Neupotz, Bad Breisig/Sinzig, and Vahnum never got beyond the blueprint stage. Even the French government—otherwise all but impervious to the anti-nuclear sentiment—backed off from building Fessenheim III and IV.

For the Rhine governments and corporations, the nuclear protests represented a major public-relations setback. Never in a century and a half had they encountered such a well-organized and effective citizen resistance to their economic development plans on the river. The prospect of turning the Upper Rhine into a second Ruhr sounded more appealing to those in the upper echelons of government and industry than it did to the farmers and villagers who lived on the Upper Rhine and equated the word "Ruhr" with urban congestion and environmental blight. Similarly, the prospect of generating vast amounts of "clean" nuclear energy sounded better when judged from afar than it did on location, where the risk of radiation poisoning loomed large. Nuclear accidents at Three Mile Island in Pennsylvania in 1979 and at Chernobyl in Ukraine in 1986 only served to underscore what many Rhine residents already felt instinctively: a nuclear plant posed a risk of unprecedented proportions, one that could not be compared to the occasional chemical spill or factory fire.

The antinuclear movement, however, represented only a minor setback to the growth plans of the Rhine's chemical and energy companies. The collective impact of hydroelectricity, petroleum, and (partial) nuclearization brought with it a quantum leap in energy generation on the High, Upper, Middle, and Delta Rhine—that is, on all those stretches that had been spared the ravages of the coal age. More energy meant more industry, more industry meant more cities, cities meant more congestion—and all of these combined meant more Rhine pollution. On the Upper Rhine alone, the new power sources translated into ten major chemical facilities, five cellulose plants, and nearly a dozen other aluminum, automobile, syn-

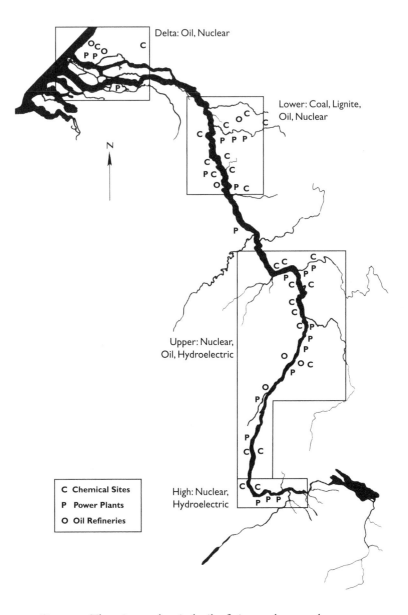

Figure 5.3 The primary chemical, oil refining, and power plant sites on the Rhine today. Various combinations of energy sources predominate on different stretches of the river. (Adapted from Cornelsen Verlag, *The Rhine*, Map 2)

thetics, electronics, and textiles factories. Urban sprawl, which in the nine-teenth century had largely been confined to Rhineland-Westphalia, now hit the Rhine-Main, the Rhine-Neckar, Alsace, and Basel regions with a vengeance. The cities of Frankfurt, Offenbach, Wiesbaden, Mainz, and Rüsselsheim melted together into one conurbation, much as had Ruhr cities a generation earlier. So too did Ludwigshafen and Mannheim, Strasbourg and Kehl, and the Karlsruhe-Ettlingen-Rastatt triangle. As these cities expanded in size and number, the river's meadowlands and forests disappeared, greatly reducing the living space of the riparian flora and fauna. There were other environmental consequences as well. Hydrodams on the High and Upper Rhine put an end to the annual salmon migration, contributing to the extinction of the Alpine salmon industry. Nuclear plants, meanwhile, began to draw such vast quantities of Rhine water to cool their reactors, and return it to the river at such an elevated temperature (up to three degrees centigrade higher), that it began to play havoc with downstream organisms and plants.[66]

Urban growth on the Upper Rhine brought another problem to the forefront: sewage disposal. Modern sanitation networks, pioneered in London in the 1830s, spread to other industrializing regions of Europe and the United States during the second half of the nineteenth century. The early systems were quite primitive, consisting mostly of "water closets" linked via a pipe or canal to a nearby river. Since most of these systems lacked purification plants, rivers that ran through major urban centers—such as the Thames and Chicago rivers—soon began to choke in human excrement. The first major conflict on the Rhine arose in 1897, when the city of Mannheim began dumping its untreated excrement into the river just eight miles upstream from where the city of Worms extracted its drinking water. After a protracted legal battle in the German court system, Mannheim was absolved of all responsibility and Worms was told to construct a costly filtration system or find a different water supply. Even the German Ministry of Health (which, given the health risks, might be expected to have sided with Worms) flatly declared in 1904 that "a river like the Rhine takes on, aside from the run-in from cities and streams, many substances originating directly or indirectly in the private household so that everyone who drinks from the apparently clear waters must take responsibility for danger to his health."[67] Mannheim's victory helped set off a chain reaction of irresponsible behavior. Basel, Strasbourg, Ludwigshafen, and many other cities felt free to dump their untreated (or partially treated) sewage into the Upper

Rhine, safe in the knowledge that there were no serious national or international consequences to fear from downstream cities. The same was true on the Middle and Lower Rhine. In 1956, water authorities in North Rhine–Westphalia discovered that there were in their state at least twenty-nine Rhine-bank cities and towns lacking adequate sewage-treatment plants.[68]

Toleration of chemical and urban pollutants during the early years of the modern chemical industry was shortsighted enough, but continued toleration after the industry had spread everywhere on the river from the Alpine heights of Grisons and Voralberg to the mudbanks of Rotterdam was a prescription for disaster. By 1925, Prussian water authorities had identified ten distinct "sacrificed stretches" on the Lower Rhine alone: at Bonn, Duisburg, Düsseldorf, and Cologne, at the mouths of the Wupper, Ruhr, Emscher, and Lippe, at the Bayer plant in Leverkusen, and at the Walsum paper mill. They were not overly concerned at the time because water quality still managed to recover between each stretch: "We did not notice any significant damage, even of a purely aesthetic nature, to the Rhine during any time of the year (low-flow periods included), and we did not notice any unpleasant sights or smells."[69] A few years later, Baden officials came up with similar results: there were distinct "sacrificed stretches" in the vicinity of all the new chemical-urban centers (Basel, Strasbourg, Karlsruhe, Speyer, Heidelberg, Ludwigshafen, and Mannheim), and at the mouths of the Neckar, Ill, and a few other tributaries. Yet like their Prussian counterparts, Baden researchers saw little indication that a full-fledged pollution crisis was brewing: in fact, most of them were still far more struck by the river's self-cleansing capacity than by the fact that the water was becoming filthier with each passing year.[70]

By the late 1930s, however, the situation had deteriorated significantly. All river water on the Lower Rhine from Bonn to the German-Dutch border was polluted to one degree or another. One of the main reasons was that because of the much greater industrial activity upstream, the river was not capable of cleansing itself before it reached the Lower Rhine. Two areas were of particular concern: the conurbation between Cologne and Duisburg, which accounted for much of the treated and untreated sewage entering the river; and the Emscher canal, which dumped the bulk of the Rhineland-Westphalia's wastewater into the Rhine just downstream from Duisburg.[71] One researcher, after recounting in detail the amount of phenol, cresol, cyanide, and other chemicals he found in the Lower Rhine between 1935 and 1937, noted: "The presence of these poisons in the river has killed off the lower rung of animals, including the Gammaridae, and caused the destruction of fish stocks."[72] Equally problematic, a study conducted in 1939 showed that fully 40 percent of Germany's total industrial effluents and 50 percent of its urban effluents were landing in the Rhine.[73] In other words, by 1939 the days were coming to an end when researchers could identify distinct sacrificed stretches. Increasingly the whole river was becoming polluted.

Most of the chemical dumpings were legal, but it scarcely mattered when they were not, since the industry had little to fear from the regulators. As late as 1969, a massive fish kill near Bingen (just downstream from the Main confluence on the Middle Rhine) led to the discovery that Hoechst routinely dumped 40 to 50 kilograms of Thiodan-related pollutants into the Main river every day in defiance of the laws.[74] A decade later, in 1979, another investigation revealed that the same Hoechst plant was bypassing its purification plant and discharging acids and acid salts into the Main untreated—this time with the full knowledge and complicity of government regulators in Hesse![75] And in 1989, two researchers discovered that for over fifty years the Swiss firms Ciba, Geigy, and Sandoz had used several sites on the Rhine near the German and French borders—at Feldreben, Lopps, and Kessler—to dispose of their chemical wastes, knowingly creating a ticking time bomb for the French and Germans who lived downstream.[76]

The massive Sandoz chemical spill of November 1986 was thus an "industrial accident" only in the narrowest sense. No one planned the fire that unleashed the chemicals that flowed through the sewage canal that led to the Rhine, but the fact that the Sandoz plant was located directly on the Rhine's banks was no accident. That its industrial sewer led straight to the Rhine channel was no accident. That its safety features were inadequate was no accident. In truth, the Sandoz spill was nothing but a "coproduct" of the shortsighted disposal methods that the chemical industry had been practicing from the beginning—methods so deeply embedded in the structures of everyday production that they could only be labeled "business as usual."

Gigantism was the defining characteristic of the Rhine chemical industry, just as it was in mining and metallurgy. As of 1988, Bayer, Hoechst, and BASF ranked first, second, and third respectively among the world's largest chemical firms, with Ciba-Geigy ranking ninth, Sandoz twenty-second, and Hoffmann–La Roche twenty-ninth. Recent acquisitions and mergers have scrambled these rankings somewhat: Sandoz and Ciba-Geigy merged to become Novartis in 1996, while Hoechst and Rhône-Poulenc (a French firm) merged to become Aventis in 1999. But these realignments have not changed the fact that the Rhine today accounts for between 10 and 20 percent of total world chemical production.[77]

The gigantism of the chemical industry spilled over into the utilities industry, creating an invisible "electropolis" (to use Thomas Hughes's term) behind much of the coal-and-chemical complex.[78] The Newcastle upon Tyne Electrical Supply Company and the Pennsylvania Power & Light Company, which emerged from the coalfields of Britain and the United States, were early examples of this electropolis. But it was really with the advent of the hydroelectricity, electrochemicals, petroleum, and nuclear fission that the energy companies turned into the megafirms they are today. Both of Germany's giants—RWE and VEBA (Vereingte Elektrizitäts- und Bergwerks)—began as coal-and-lignite enterprises in the Ruhr and Ville before branching into other energy sectors, notably petroleum and nuclear. In France, L'Energie Electrique du Rhin (now part of Electricité de France) began as a hydroelectric company, then moved in a big way into the nuclear industry. The same transformation occurred in Switzerland with Nordost-schweizerische Kraftwerke, the dominant energy company on the Swiss Rhine. In the Netherlands, it was mostly petroleum-based companies such as NV ENECO and Dutch Shell (which merged into ENECO Shell Energy in 1999) that arose side by side with that country's Rhine-based energy needs.

The interconnectivity of the Rhine coal, chemical, hydroelectric, petroleum, and nuclear industries lent them greater political leverage than any of them would have had on their own. They used this clout to defeat the wastewater clause in the 1913 Prussian Water Law, to shape and control the Wupper Cooperative when it was finally established in 1930, to stymie the creation of other cooperatives, to thwart international regulations, and (above all) to propagate the "sacrificed stretches" principle long after it had been utterly discredited in practice. During the nineteenth century the chemically burdened stretches were largely confined to the Lower Rhine and its tributaries, many of which were already suffering from the ill-effects of mining and metallurgy. Hydroelectricity, however, brought chemicals to High and Upper Rhine, and even (to a limited extent) to the headwaters of the Alpenrhein and Aare. After 1945, petrochemicals and nuclear power extended the sacrificed stretches to the Middle Rhine and Delta Rhine while at the same time multiplying the burdens everywhere else on the river.

One small feeder stream after another was poisoned, one tributary after another burdened, one Rhine section after another colonized—until the day finally arrived when the entire Rhine basin had been turned into one massive coal-and-chemical warehouse, and the river itself had become one long "sacrificed stretch."

It is said that a frog, when placed in a cold pot of water on a heated stove, will gradually boil to death because its body acclimates to the warming temperature instead of registering danger. Like the Emscher, Ruhr, Lippe, Erft, and Wupper before it, the Rhine made the transition from a clean-flowing to a near-dead river so incrementally that its plight was overlooked, at least until the 1970s when its water quality had reached such a low point that only the most blinkered could fail to see.

A salmon swam into the Rhine
And headed for the Alps
Springing over rapids
And leaping over falls
Until one day along the way
It slammed against a wall.

— Christian Morgenstern (1910)

Too much industry
No fish in the Rhine
Lorelei poisoned
Too much embarrassment.

— Allen Ginsberg (1979)

CHAPTER 6

Biodiversity Lost

The Rhine made a cameo appearance in Mary Shelley's Gothic tale, *Frankenstein; or, the Modern Prometheus*, published in 1818. Dashing from Switzerland to Britain with a desperate scheme to create a bride for his monster, Victor Frankenstein decided to sail down the Rhine to Rotterdam and from there catch a sea vessel to London. His route, unfortunately, was anything but speedy. He consumed a full five days sailing from Strasbourg to Mainz, including an idle day in Mannheim. He got sidetracked again in Mainz, an obligatory resting spot for all real and would-be Romantics ("This part of the Rhine, indeed, presents a singularly variegated landscape," wrote Shelley in Radcliffe-like rapture. "In one spot you view rugged hills, ruined castles overlooking tremendous precipices, with the dark Rhine rushing beneath; and, on the sudden turn of a promontory, flourishing vineyards, with green sloping banks, and a meandering river, and populous towns, occupy the scene"). Contrary winds and a sluggish current finally convinced him to abandon the Rhine altogether at Cologne, and he took a land route to Rotterdam instead. So slow was Frankenstein's progress that his monster was able to shadow him on foot. "I left Switzerland with you," the monster exclaimed when they met again in Britain. "I crept along the shores of the Rhine, among its willow islands, and over the summits of its hills."[1]

Had Frankenstein attempted the same route a few decades later, he would have found it a much quicker and more pleasant journey (and his monster would have found fewer willows and islands to hide amidst). The Netherlands Steamboat Company began regular passenger service between Rotterdam and Cologne in 1825. Two years later, the Prussian Steamboat Company (predecessor to the now ubiquitous Köln-Düsseldorfer) started serving the Middle Rhine from Cologne to Mainz.[2] The Prussian firm alone carried 18,000 passengers in its inaugural year, fully half of whom were British subjects in search of a Romantic holiday. The number of passengers surged to 685,000 in 1847, 1.1 million in 1867, and 1.9 million in 1913, despite competition from a dozen other steamer lines.[3]

Railroads, meanwhile, began to provide tourists with an additional mode of transportation, especially on the Upper Rhine where travel by ship remained unpredictable and precarious well into the twentieth century. The most important were the Alsatian Line from Basel to Strasbourg (1841), the Main-Neckar Line from Heidelberg to Frankfurt (1846), and the Baden Line from Basel to Mannheim (1849). By 1860 most of the left bank was lined with train tracks, and by 1870 most of the right bank as well. Cycling tours began on the Rhine in 1902, once the old towpaths were converted into bike routes. For the well-to-do, automobile roads were added in 1909. Finding the best tourist sites and most comfortable hotels also became easier. Whereas Shelley had to rely on the impressions of Radcliffe and other Romantics, visitors after 1828 could turn to J. A. Klein's pathbreaking guidebook, *A Trip on the Rhine from Mainz to Cologne*, aptly subtitled "A Handbook for Speedy Travelers." Renamed the *Baedeker Guide* in 1849, it has been the mainstay of Rhine tourists ever since.[4]

The ever-quickening pace of travel would have astonished the Congress of Vienna diplomats. Nobody could have foreseen in 1815 that the Rhine would one day be overrun with roads, bridges, rail lines, bike paths, hotels, restaurants, and camera-toting tourists. Yet the diplomats would have been equally astonished at the monstrous disfigurement of the riverbed and the denuding of its alluvial forests. In 1815, the river was full of islands and braids, its banks a continuous corridor of lowland forests and meadows. Its catchment area supported innumerable plant and animal species, its trees a vibrant bird population, its waters a cornucopia of fish. By 1975, the Rhine had lost most of its forests and floodplain; it supported only a tiny fraction of its original flora and fauna; and it had earned the nickname "Europe's romantic sewer" and "Europe's sewage dump." One hundred fifty years of hydraulic tinkering had turned the Rhine into a soulless shadow of its former self. Once clean, it was now filthy; once broad, it was now narrow; once bursting with life, it was now half-dead.

Water Pollution

Researchers rely on two different but related methods for measuring the quality of a river's water: biological analysis and nonbiological analysis.

Biological analysis consists of periodic saprobic water sampling based on a four-point sliding scale. Class I ("unpolluted") signifies that the water is oxygen-rich and free of chemical contaminants and thus provides an optimum environment for riverine life. Class II ("moderately polluted") means that water quality is compromised but that oxygen levels are still normal enough to support many life forms. Class III ("strongly polluted") connotes the presence of large amounts of harmful substances, high bacteria levels, the near-disappearance of higher plant life, and the spread of pollutant-resistant sponges, leeches, and ciliates. Class IV ("completely polluted") means the water is so oxygen-depleted (eutrophic), usually as a consequence of being supplied with excessive phosphorus or nitrogen-based nutrients, that it supports bacteria, flagellates, and ciliates, but not most other life forms (see figure 6.1).[5]

Biological analysis provides a useful gauge of a river's overall water quality, but nonbiological analysis is better for identifying the specific causes of

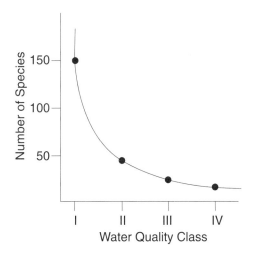

Figure 6.1 The relationship between biodiversity and water quality. Note the precipitous drop in biodiversity between Class I ("unpolluted") and Class II ("moderately polluted"). Most river organisms cannot adapt even to moderate amounts of industrial-urban pollutants. (Source: Brenner, Hantge, and Kinzelbach, *Wie sauber ist der Rhein wirklich?* 138)

degradation. For rivers, there are three critical nonbiological pollution categories: thermal, chloride, and (most important) chemical. Thermal pollution refers to all human-induced changes in the river's water temperature. In practice, it almost always means that power plants have artificially raised the temperature by dumping heated wastewater from their cooling facilities into the river. An alteration in temperature is not toxic per se, but it does affect the number and composition of species capable of living in the streambed. Chlorides are common salts that enter the river through a variety of agricultural and industrial activities. They are not toxic to humans and cause no harm to marine organisms once they reach the sea. High chloride levels, however, give river water a brackish quality, thereby diminishing the number of freshwater organisms capable of living in the bed and channel.

Chemicals—the third category—are the primary pollutants on the Rhine, as they are on all of the world's major industrial rivers. Hundreds of different chemicals find their way into rivers every day, but three groups cause the greatest concern: heavy metals (zinc, copper, chromium, lead, cadmium, mercury, and arsenic), which enter rivers primarily at industrial and mining sites; chlorinated hydrocarbons, such as polychlorinated biphenyl (PCB) and, until it was banned, dichlorodiphenyltrichloroethane (DDT); and phosphate-based and nitrogen-based substances, chiefly fertilizers and detergents. The first two groups—heavy metals and chlorinated hydrocarbons—are of special concern because they are persistent chemicals that bioaccumulate: they pass unmetabolized from simple organisms to more complex organisms, magnifying in concentration as they move up the food chain. The third group—phosphorus and nitrogen—poses a different set of problems. These substances act as inorganic nutrients when introduced into a lake or river, spurring the growth of oxygen-depleting phytoplankton (free-floating algae). Excess nutrients can result in the gradual eutrophication of a river, with negative consequences for fish and other organisms that depend on dissolved-oxygen for respiration. In the Rhine basin, this problem is mostly confined to three areas where the current is too sluggish to flush phytoplankton downstream on its own: at Lake Constance (at the Austrian-Swiss-German border in the Alps), on the Lower Rhine in North Rhine–Westphalia (the Ruhr), and along the coastal estuaries of the Dutch delta.

Data collected in the 1960s and 1970s by an international team of water researchers using uniform methods demonstrated an ongoing deterioration in water quality on all stretches of the river from headwater to mouth. By 1970, for instance, the Rhine's chloride load had reached an unprecedented 365 kilograms per second at the Dutch-German border. Only about one-fourth of this salt (93 kilograms) was derived from natural sources, mostly from rocks in the river's basin; the remaining three-fourths (272 kilograms) came from anthropogenic sources. The Alsatian potash plants, clustered around the silvinite mines near Mulhouse on the Upper Rhine, accounted for nearly half of the human-induced salt load: 130 kilograms per second, amounting to 7.5 million metric tons per year. Most of the river's chloride load came from sodium chloride, a waste product that results when silvinite is transformed into potassium chloride for use as a fertilizer. The Mosel tributary was the second biggest contributor to the Rhine's chloride content, with the salts coming primarily from a variety of French and Luxembourg sources. The remainder of the river's chloride load came from soda works and mine water from the Ruhr region, and from agricultural runoff all along the river. The annual salt load was at least one-third higher than maximum acceptable levels.[6]

Also, by the 1970s the accumulation of heavy metals in the Delta Rhine's silt had reached the point that it far exceeded safe limits for zinc, copper, chromium, lead, cadmium, mercury, and arsenic. Flocculation and filtration methods at sewage treatment plants were capable of removing the bulk of these contaminants from the drinking water supply, so they posed little danger to human consumers. But the river's silt had become too contaminated to be used for land reclamation, a major enterprise in the Netherlands. Moreover, every time the Rhine flooded, this silt was deposited onto the nearby fields, rendering them unsuitable for farming and grazing. The content of copper in the Netherlands' floodplain, for instance, stood at 160 parts per million (ppm) in 1974, over six times the level considered safe for arable soils (sheep, for instance, are so sensitive to chronic copper toxicity that grass containing as little as 20 ppm is enough to kill them). Mercury and cadmium contents in the soil were twenty-five times higher than levels considered safe. Zinc levels were eleven times higher, lead levels seven times higher, arsenic levels five times higher, and chromium levels four times higher than acceptable limits. Researchers also found that the Rhine was carrying 15 metric tons of nonbiodegradable organic substances every day, chiefly chlorinated hydrocarbons from pesticides and industrial processes. These substances passed through the biological sewage treatment plants without being removed from the water.[7]

What this added up to was a grim tale of water degradation. By 1975, the year generally considered to be the peak year of Rhine pollution, most of the Upper and Middle Rhine belonged to Class II or Class III ("moderately" to "strongly" polluted), while the Lower Rhine stood at Class III and Class IV ("strongly" to "completely" polluted). The entire navigable Rhine, in other words, was so polluted that all natural biological conditions had been compromised, with the level of degradation increasing as the water moved downstream. A special German commission, set up in the 1970s to assess Rhine water quality between Basel and Rotterdam, concluded that the entire navigable part of the river had become so contaminated with industrial and urban pollutants that it had been transmogrified into one long sacrificed stretch.[8] Similarly, a study undertaken by a consortium of Rhine waterworks in the 1970s determined that fully half of the river's pollutants came from just six sources, all of them tied to chemical production: a cellulose factory in Strasbourg, a cellulose factory in Mannheim, an Alsatian potash mine, the Bayer chemical plant in Leverkusen, and the urban centers of Basel and Strasbourg.[9] Finally, in 1985 (just a year before the Sandoz chemical spill), Dutch waterworks researchers identified the following substances in Rhine water at the Dutch-German border: 11 million metric tons of chlorides (waste salts), 4.6 million tons of sulfates, 828,000 tons of nitrates, 284,000 tons of organic carbon compounds, 90,000 tons of iron, 38,200 tons of ammonia, 28,400 tons of phosphorus, 4350 tons of zinc, 2500 tons of organic chlorine compounds, 681 tons of copper, 665 tons of lead, 578 tons of chromium, 530 tons of nickel, 126 tons of arsenic, 13 tons of cadmium, and 6 tons of mercury.[10]

The Rhine Floodplain

"Far from being the gale which blew away the desiccated feudal leaves," wrote the historian T. C. W. Blanning, "the French Revolution is better likened to a chainsaw, which felled an ancient, gnarled, but still flourishing oak."[11] By "oak," Blanning meant the Holy Roman Empire, the vast polyglot Habsburg realm that towered over Germany until it was toppled by the Napoleonic war machine. But his metaphor can be taken literally as well: the French Revolution cleared the path for the Rhine Commission and its army of engineers, who engaged in a century-long campaign against the river's floodplain. As the floodplain began to disappear, so too did the river's once-flourishing stands of oak, elm, willow, ash, poplar, alder, and beech trees. And once the trees disappeared, so too did many of the life forms that depended on the floodplain for survival.

Virtually all of the Rhine's lowland forests, meadows, and marshes have been expunged over the past two hundred years. In 1815, floodplain enveloped around 2300 square kilometers of the river (even more if the Alpenrhein and Aare tributaries are included in the totals). This floodplain once formed a single continuous river corridor, varying in width from a few hundred meters to fifteen kilometers and stretching over a thousand kilometers from Lake Constance to the delta mouth. By 1975, most of this land had been taken over by farms, pastures, businesses, train tracks, roads, and cities. The High Rhine lost 86 percent of its floodplain, the Upper Rhine 87 percent (see figure 6.2), the Middle Rhine 96 percent, and the Lower Rhine a whopping 98 percent. Less than 500 of the original 2300 square kilometers is still subject to periodic flooding—and even most of that 500 square kilometers is now part of the human-built environment.[12]

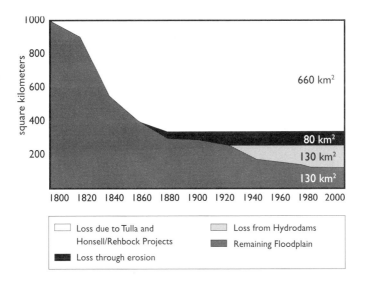

Figure 6.2 Floodplain loss on the Upper Rhine between 1800 and 1996. River engineering has all but obliterated the Rift valley floodplain at the base of the Alps. (Source: ICPR, *Bestandaufnahme*, 26)

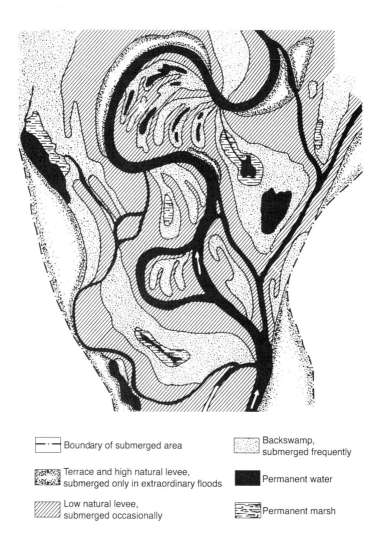

⊢·⊣ Boundary of submerged area	Backswamp, submerged frequently
Terrace and high natural levee, submerged only in extraordinary floods	Permanent water
Low natural levee, submerged occasionally	Permanent marsh

Figure 6.3 Features of a typical floodplain. A river provides many diverse watery environments for aquatic and semiaquatic species. (Adapted from Welcomme, *Fisheries Ecology*, 6)

It was the forests, meadows, marshes, and reeds of the floodplain that once gave the Rhine its geographic breadth, biological diversity, and ecological dynamism (see figure 6.3). Only fish and a handful of other organisms spend their entire life in the river itself, and even they are highly dependent on the surrounding nonaquatic environment for organic nourishment. Most riparian organisms cling to the channel bed or bank, or dwell in the wetlands, grasslands, hills, trees, bushes, meadows, and valleys around the flowing water. Wetlands are especially important because they serve as nursery grounds for fish, shellfish, aquatic birds, and other animals. They also remove excess phosphorus, nitrogen, and other pollutants from the river. The roots of trees and bushes help stabilize the banks and even out the flow of water during drought and flood periods. Foliage provides shelter, shade, and nesting spots for birds and other animals. Thus, when the Rhine lost most of its floodplain it also lost most of the living space upon which its biodiversity depended. Some species went extinct, while others just went away. A handful of species were able to take advantage of abandoned or new ecological-niches. But no part of the riparian population remained unaffected by the destruction and changes.

Where once there was a continuous river corridor, there now are just a few natural or near-natural stretches, tiny dots on a landscape otherwise dominated by human enterprise. That any of these patches managed to survive long enough to get official protection in the 1960s and 1970s was more a matter of luck than planning: usually it just meant that no city or industry had yet reached that spot, or that the cost of reclamation exceeded the prospect for development. The Upper Rhine has the highest number of protected areas: Kühkopf-Knoblochsaue (2378 hectares), Taubergiessen (1600 hectares), Rastatter Rheinaue (845 hectares), Hördter Rheinaue (818 hectares), Lampertheimer Altrhein (516 hectares), Ketscher Rheininsel (490 hectares), and Mariannenaue-Fulderaue-Illmenaue-Rüdesheimer Aue (470 hectares). The protected areas of the Middle Rhine are mostly clustered around the Seven Mountains region south of Bonn, where efforts to preserve the famous Drachenfels castle in the early nineteenth century had the effect of preserving much of surrounding landscape in the process. The Lower Rhine still has two medium-sized patches—Rheinaue Walsum (345 hectares) and Bienen-Praest (340 hectares)—as well as several small ones. There are also other remnant patches on the river, ranging from a tiny two hectares to a modest-sized two hundred.[13] By way of comparison: the Tulla

Project opened up 10,000 hectares of former floodplain for agricultural and industrial use on the Upper Rhine, and that was by no means the largest reclamation project on the river.[14]

Because these natural and near-natural spaces are far too puny to function as a continuous river corridor, most species find their living spaces and migration routes greatly constricted. Nonetheless, these patches harbor almost all that remains of the Rhine's native vegetation and old-growth forestland. One of the most spectacular of them — the Taubergiessen — contains dense stands of oak, ash, elm, poplar, mountain maple, and linden trees, as well as a multitude of willows (including white, almond, osier, purple, and crack). Ivy, clematis, grapevine, woody nightshade, and other vines can be found there, as can hazel, dogberry, blackthorn, hawthorn, spindleberry, guelder rose, honeysuckle, and related bush varieties. Hugging the ground are a wide variety of native shrubs, herbs, and mosses, and on the water's edge are some of the few remaining original reed species. The density of trees, brush, and undergrowth provides a haven for insects. Nearly 40 percent of Central Europe's butterfly species are found in Taubergiessen, as are 80 percent of its dragonfly species. Also at home there are nearly 1000 beetle and 350 hymenopteron (wasps, ants, bees) species, as well as many plants and animals on the "Red List" of endangered species.[15]

Many migratory and nesting birds have been able to maintain a presence on the Rhine by utilizing the few remaining old-growth forests and mead-

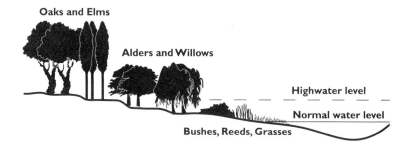

Figure 6.4 Profile of a natural riverbank. A riparian corridor provides the vegetative habitat upon which birds and other animals depend for their survival. (Adapted from Solmsdorf, Lohmeyer, and Mrass, *Ermittlung und Untersuchung*, 34)

owlands of Taubergiessen and similar preserves. The Upper Rhine's remnant oak-elm stands, for instance, contain as many as 285 brooding pairs per ten hectares, among the highest bird densities in all of Central Europe.[16] Songbirds have found a sanctuary in the Erpeler Ley and Langenbergskopf preserves of the Middle Rhine, where a small patch of original Pontic-Mediterranean flora has survived intact. Similarly, the Lower Rhine's remaining floodplain regions (the Xantener Altrhein, Millinger Meer, and Hurler Meer) have managed to sustain a wide variety of marsh birds.[17] But bird density is not an inherent sign of health: it often just means that birds have little choice but to feed and nest in close proximity to one another on the region's few remaining trees, bushes, and marshes.

Moreover, these patches are neither geographically diverse nor large enough to preserve anything approaching the Rhine's entire array of native bird species. Forest destruction has reduced the habitat of woodpeckers and other birds that depend on tree holes for nesting. As the numbers of shrubs, bushes, and thickets declined, so have the numbers of warblers, thrushes, tits, buntings, and finches. Birds that nest on gravel and sandbanks—the plover, sandpiper, and common tern—have become rare or disappeared from the river. Marsh drainage has endangered the survival of storks, heron, bittern, owls, and wading birds. In addition, a few birds have been hunted to death, either for sport or for crop protection, notably the short-toed eagle, common raven, peregrine falcon, and capercaillie. And all species have been affected by the use of agricultural pesticides and chemicals, and by the hustle and bustle of urban life, from which there is no sanctuary.[18]

Because they are highly mobile, birds have generally been able to adapt to the new riverscape better than most other river species, though many of them are now forced to use secondary biotopes rather than their primary ones for nesting and feeding. The altered conditions have, in some cases, even created new ecological niches (watery areas at the lignite mines and at dredging sites, for instance) that have attracted new bird species to the river. But bird populations have dropped considerably in absolute numbers over the past century, and so has the number of species still finding the river suitable as a habitat. No longer seen in the Lower Rhine's marshlands are the dunlin, black grouse, roller, wood sandpiper, ortolan bunting, and eighteen other bird varieties. Endangered there are all diving birds, herons, songbirds, and terns, as well as most ducks, geese, falcons, rails, and wading birds. Twenty-three bird species can no longer be found anywhere between Lake Constance and Bingen, including the night heron, black stork, pintail duck, goosander, lesser spotted eagle, stone-curlew, great bustard, red-legged partridge, and osprey. Another twenty-three species—the sand martin, red kite, marsh harrier, quail, lapwing, eagle owl, and bluethroat, among them—have declined significantly in numbers. Of the 125 nesting species found in the German right-bank state of Baden-Württemberg, two-thirds are considered endangered. In almost every instance, scarcity of habitat accounts for their endangerment: a scant 1.3 percent of the state's entire territory is still suitable for bird nesting.[19]

The Rhine's mammal populations have also been greatly impacted, especially its semiaquatic species (or "river mammals"). The beaver, despised by fisheries and foresters alike, was driven from the Rhine basin as early as the 1830s. The otter was nearly wiped out a century later through the combined onslaught of hunting, habitat destruction, artificial banks, and pollutants (especially mercury and hydrocarbons). Under protection since 1943, its habitat is confined today to the delta region. Meanwhile, several nonnative mammals—the muskrat, coypu, and raccoon—have taken over the niches once occupied by their native brethren, in some cases with ecologically disruptive results. The muskrat, for instance, was introduced to the Rhine from the United States in 1905 in the false hope that pelt hunters would keep its numbers in check. Lacking natural enemies, and highly tolerant of river pollution and concrete banks, its numbers exploded to such an extent that control measures had to be instituted in the 1930s. Other mammal populations have been impacted as well. Bat populations ("flying mammals") declined precipitously on the river when deforestation deprived them of resting and sleeping spots. Several terrestrial mammals, notably mice, have also found their living space greatly constricted by urban and industrial development.[20]

Amphibian and reptile numbers have also steadily declined. Once bountiful everywhere on the river, they began to vanish as the natural spaces on the water's edge evaporated. Remnant populations are still extant on most stretches of the river, but three species are considered especially endangered: the tree frog, dice snake, and adder. The tree frog, once plentiful on the Upper Rhine, has suffered from a dearth of still-water areas for its larvae and a reduction in the number of trees and foliage. The dice snake, Europe's most aquatic species, requires clear-flowing water, river shallows, and sunning spots on trees overlooking the water's edge—as well as a large supply of small fish and amphibians to eat. It does not adapt well to water pollution and canalized banks and has thus lost most of its former habitat on the Middle Rhine, Lahn, and Ahr. The adder feasts on small mammals and lizards; once plentiful on the Lower Rhine, it became rare when its former hunting ground was turned into farmland and pasture.[21]

The Rhine's invertebrate populations have also been affected by engineering and the loss of riparian vegetation—though the record is far from clear because the first systematic investigations took place in the early twentieth century, by which time the negative impact of river engineering was already being felt. The survival of any particular native species, and the success of any nonnative one, depended on whether changes in the river's bed, bank, and floodplain increased or decreased its habitat, and whether water pollution augmented or reduced its food supply. The so-called lithophiles—organisms that adhere to hard surfaces, such as freshwater limpets, snails, leeches, sponges, moss animals (Bryozoa), and hydroids (Coelenterata)—have generally prospered in the new Rhine environment, for obvious reasons. Previously confined to the canyon walls of the High and Middle Rhine, many native lithophiles have been able to spread via the new rock-lined embankments to nearly every nook and cranny of the river. Several nonnative ones have also crawled to the Rhine via canals and embankments, or hitchhiked there on the hulls of ships. The number of leech species, for instance, grew from four in 1900 to twelve in 1980, while the number of moss animals rose from four to nine during the same period. Sponges benefited as well. All have expanded at the expense of those organisms that depend on a soft rather than hard surface.[22]

The number of crustacean species (crabs and shrimp) rose from three to thirteen, and the number of mollusks (clams and snails) from twenty to thirty-three since the early twentieth century. But here, too, the apparent increase in biodiversity masks a more troubling development. Several native species—including the age-old Rhine mussel—have disappeared entirely from the new river. The successful transplants, meanwhile, have all been species that can tolerate saltier, warmer, and eutrophic (deoxygenated) river water.[23] The most notable newcomer is the zebra mussel, a coin-sized black-and-white striped mollusk that has become the bane of many industrial rivers in North America and Europe. It first arrived on the Rhine in 1835, probably from the Caspian Sea. Highly tolerant of salt, and a voracious consumer of phytoplankton, the zebra mussel has come to dominate the Rhine's benthic (bottom-dwelling) community by sheer force of numbers—estimated at 700 to 800 million adults between Basel and Rotterdam. Other newcomers—the freshwater crayfish, shrimp, beach flea, river snail, and bladder snail among them—are all physiologically, morphologically, and behaviorally adapted to industrial pollutants, dissolved salts, and other harsh river conditions.[24]

"River insects" (beetles, bugs, flies, and dragonflies) have also been affected by Rhine engineering, though the exact impact is obscured by the absence of clear-cut data from the past. Most insects utilize the streambed itself only during the larval stage, but even the nonaquatic phases of their life cycle keep them mostly within the riparian zone. Many are still found in great variety on the Rhine, including dragonflies (around fifty species) and beetles (over a thousand species). However, three insects crucial to the Rhine's food chain—the mayfly, stonefly, and caddisfly—have fallen drastically in the past century. All three are highly dependent on unpolluted and oxygen-rich water and are therefore good indicators of river quality. Their taxa numbers, which stood at 111 in 1900, fell to just three in 1971. Taxa richness rose again as water quality began to improve after 1975, but as recently as 1980 there were still only 46 taxa found on the Rhine, less than half the original number.[25]

Rhine Fish

In 1880, the river was home to 47 fish species, of which 45 could be found in the High, Upper, and Middle Rhine, and 42 in the Lower and Delta Rhine. By 1975, only 23 native species were still commonly found in the upper stretches, the rest having disappeared or become so rare they were endangered. The Lower and Delta Rhine's numbers, meanwhile, dropped to 24 in the same time span. Improvements in water quality after 1975 reversed the downward spiral, and by 1986 a total of 38 of the original 47 native species swam once again in the Rhine. Yet many of these species continue to be endangered, their habitat restricted to a few patches of the river left untouched by engineering (typically the Upper Rhine's remnant bed, which flows parallel to the Grand Canal d'Alsace). Many other fish species remain plentiful only because they are sustained through artificial propagation in hatcheries.[26]

As fish counts dwindled and species disappeared, anglers and biologists tried introducing nonnative species, some of which are still found in the river. The Danube zander, introduced in 1822, managed to establish itself permanently after intensive fish-stocking efforts were undertaken a century later. The North American rainbow trout, introduced in 1881, spread quickly, though it has remained dependent on hatcheries for reproduction. The Danube brook trout, introduced in 1889, adapted easily to its new Alpine home, not least because of its high tolerance for acidic water; but it spread at the expense of the native Rhine brook trout, with which it competes for food, shelter, and breeding grounds. Both the black bullhead and pumpkin-seed sunfish, introduced in the late nineteenth century, have found a permanent niche in the river, without ever becoming numerous. Efforts to transplant the Danube huchen as an ersatz salmon failed in the mid-twentieth century. Most recently, three Far Eastern carp species have been transplanted to reduce the amount of unwanted macrophytes (rooted aquatic plants) and phytoplankton in the water, but all three species are sustained solely through pisciculture.[27] (The practice of introducing exotic fauna has been dubbed the "Frankenstein effect" because typically the long-term damage it does to native communities far outweighs the short-term benefits it brings to fishermen.)[28]

The survival of any given Rhine species, native or not, depended on whether it could tolerate polluted water, adapt to a deeper bed, swim in a swifter current, hide from predators, and find suitable nourishment. But one issue towered above all others: the presence or absence of adequate spawning and nursing grounds. Species dependent on aquatic plants, branches, roots, gravel, or sand to hide their eggs have had difficulty finding suitable sites in the new riverscape. So have species that require slow-moving braids, branches, backwaters, riffles, or deep-water pools for reproduction. The "universalists" (species that can adapt to a variety of spawning conditions) have generally been able to utilize secondary and tertiary spots as reproduction sites. But "specialists" (species that require precise and stable breeding sites, such as the bitterling, which deposits its eggs in the gills of freshwater clams) have had a much more difficult time surviving. Many nonpredacious species—those that rely on plants, insects, and small organisms for nourishment—have managed to eke out an existence, however meager, in the new river. But predatory species, such as the pike, zander, chub, burbot, wels, and perch, all of which nourish themselves on other fish, have had to cope with an ever-dwindling food supply. Finally, migratory species have fared worse than stationary ones because their life cycles subject them to many stretches of the river and therefore to all of its ills.[29]

The figures tell the tale. Salmon, shad, and sturgeon—for centuries the most celebrated fish on the Rhine—disappeared completely from the river. Other migratory species—the lampern, sea lamprey, North Sea houting, and sea trout—became rare. Most numerous today, by far, are three fish from the Cyprinidae (carp) family: the roach (36.1 percent), bleak (25.6 percent), and common bream (11.8 percent). Tolerant of polluted and degraded streams, they now account for 73.5 percent of all fish swimming between Basel and Rotterdam. The next six most common species are the eel (6.8 percent), dace (6.8 percent), perch (4.3 percent), chub (2.3 percent), white bream (1.6 percent), and gudgeon (1.4 percent), which collectively account for another 23.2 percent of the river's fish. The remaining extant species—grayling, smelt, pike, mudminnow, ide, barbel, bitterling, minnow, rapfen, souffie, loach, burbot, stickelback, bullhead, and flounder, among others—are now found in such small numbers that they collectively account for a mere 3.3 percent of the fish population.[30]

The Rhine's Alpine regions (the High Rhine, Lake Constance, Alpenrhein, and Aare) have been less affected, primarily because they all lie upstream from Rheinfelden, the endpoint of Rhine navigation. Native to these tributaries and stretches are fish that prefer cold, swift, and oxygen-rich waters, with trout and grayling as the predominant species. The High Rhine supports barbel, brook trout, bullhead, minnow, cod, and (rarely) loach, much as it did before; but salmon have disappeared and roach have increased, both as a consequence of downstream engineering. Lake Constance still has a bountiful supply of freshwater houting, perch, pike, common bream, grayling, and lake trout, not least because fish hatcheries watch over the stocks on behalf of local fishermen. The Alpenrhein also supports a wide variety of fish: the native brook trout, grayling, minnow, bullhead, barbel, nase, and souffie, as well as the nonnative Danube brook trout and American rainbow trout. Water pollution, river engineering, and fish management have nonetheless affected both the number and composition of the populations now found in the Alpine regions. In 1930, freshwater houting accounted for 75 percent of all fish caught in Lake Constance, but by 1979 that number had dropped to 37 percent. Similarly, grayling and lake trout once migrated from Lake Constance up the Alpenrhein to the Vorderrhein and Hinterrhein, but since 1956 the Reichenau hydrodam has blocked that route.[31]

For Rhine biologists, the most devastating consequences of river engineering were the disappearance of native species, the destruction of spawning and feeding grounds, the relative change in fish composition, and the absolute reduction of fish numbers. For Rhine fishermen, it was the loss of three major commercial species—sturgeon, shad, and salmon—none of which could survive in the new river. Sturgeon are native to the coastlines and rivers of Europe from the Baltic to the Black Sea. The Rhine and its tributaries were once part of their migratory routes and spawning grounds. Sturgeon were never plentiful enough to play a significant role in Rhine commerce, but their enormous length (up to three meters) and their lucrative eggs (caviar) made them a highly prized catch. Decreases in spawning grounds and delta habitat, increases in water pollution, and overfishing all conspired against the sturgeon, and by the early twentieth century they had begun avoiding the river altogether. In 1893, 832 sturgeon were caught in the Rhine delta. That number dropped to 462 by 1903 and to a meager 28 by 1913.[32]

The allis shad is better known on the Rhine by its colloquial name "mayfish." A herring-like species, it spent most of its life in salt water, then migrated upriver in May (hence the name) to breed. Allis shad found spawning grounds in the upper stretches of the Rhine and its tributaries as high as the Laufenburg rapids in Switzerland. Once a mainstay of the spring and summer fishing season, the allis shad's numbers plunged precipitously in the late nineteenth century. Overfishing was the main cause. Anticipating a crisis, the riparian states signed the International Mayfish Convention in 1895. Designed to revive the stocks through hatcheries, it failed because it imposed no catch limits on mature shad. Already in 1900 only a few allis shad were seen on the upper reaches of the river. The Mosel too, once a major shad fishing site, was showing signs of exhaustion. The delta catch, meanwhile, peaked at 310,000 in 1885, then dropped to 76,000 in 1905, and then to a paltry 190 in 1919. By the 1930s, the once-famous May migrations no longer attracted the attention of fishermen. The term mayfish is often applied to the closely related but smaller fintis shad, which also migrates from the sea but remains in the brackish waters of the delta and the lower stretches of the river to propagate. Commercial fishing of the fintis shad remained a prosperous enterprise until the 1950s, when their numbers also suddenly plummeted, leaving only a remnant population today. Their demise coincided with an upturn in delta engineering projects, which blocked their migratory routes and deprived them of spawning spots.[33]

Most fish species disappeared from the Rhine with little or no fanfare, bemoaned by a few specialists or fishermen but otherwise consigned to

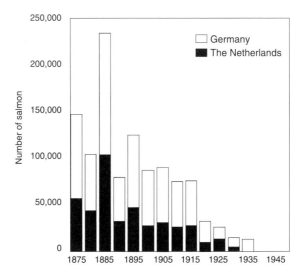

Figure 6.5 German and Dutch salmon catches, 1875–1950 (Adapted from Schulte-Wülwer-Leidig, "Ecological Master Plan," in Harper and Ferguson, eds., *The Ecological Basis for River Management*, 506)

oblivion as much by human neglect as anything else. That was not the case with salmon. Salmon were the embodiment of the river's soul and the food-stuff of myth and legend. Their utter destruction in the twentieth century generated much political debate, diplomatic negotiation, and public hand-wringing—albeit to no avail. The salmon catch peaked at around 225,000 in 1885, then dropped to under 100,000 by 1900. By 1935 the annual catch had fallen below 15,000, and by the 1950s salmon and the salmon industry had disappeared completely (see figure 6.5).[34]

Rhine salmon typically spawned between September and December, mostly in the cool-water gravel beds of the river's tributaries but also in the braids and backwaters of the Upper Rhine. After a year or more, the juveniles descended to the sea to mature, and then returned to the river a few years later to spawn the next generation of salmon. Guided by their sense of smell, they migrated back to exactly where they had originated. Although all Rhine salmon belonged to the same species (*Salmo salar*), fishermen typically categorized them based on season and size. "Winter salmon" were the largest, averaging ten or more kilograms in weight. They ascended the river between November and May, with the greatest numbers coming in March. Their journey was leisurely, taking as long as twelve to fifteen months. Mostly male and still sexually immature, they would not mate until the following winter. "Small summer salmon," weighing five to eight kilograms, were most numerous in June and July. They were near-mature fish, predominantly females in their third or fourth year. They swam directly to their spawning grounds, which they reached within three to eight weeks. The third group, "Saint Jacob salmon," were so named because their arrival coincided with the celebration of Saint Jacob's Day on July 25. They were almost entirely males in their third year, weighing three to six kilograms. The fourth group, "large summer salmon," were fully mature, egg-laden females en route to spawn. They were most numerous from September to December.[35]

The riparian states, seeing a crisis looming, signed the Salmon Convention in 1885, the same year that the catch peaked. Designed to protect the stocks during the spawning season, the treaty forbade use of dragnets in the delta region between August 16 and October 15, and elsewhere between August 27 and October 26. It also prohibited fishing devices that extended more than halfway into the river, and put a halt to Sunday fishing. Further, it required member states to establish hatcheries, and obligated fishermen to turn over the roe and milt of all mature salmon caught between October 15 and December 31 for use in pisciculture.[36]

Despite its elaborate regimen, the treaty did nothing to slow the downward spiral, let alone replenish the salmon stocks. It failed in large part because the riparian governments did not address two of the most important destructive forces at work on the river: dredgers and dams. Dredging was an integral part of all river rectification work. Engineers could not alter tributary mouths, reinforce banks, deepen harbors, or create uniform shipping lanes without removing a considerable amount of the river's sand and gravel deposits. (Some of this sand and gravel was used by the Germans and French for their trenches and fortifications in both world wars.) Dams were equally important to engineers, especially on the High and Upper Rhine, home to the salmon's best spawning grounds. They were used for flood control, for low-water regulation, and (after 1898) for hydroelectric production. For a salmon treaty to be effective it would have had to impose restrictions on navigation and industry, which none of the riparian states were willing to permit, collectively or individually.

At the time, the German government suspected that one of the main reasons for the demise of the salmon was overfishing at the Rhine mouth. In 1897, for instance, a member of the German Reichstag, Freiherr Heyl zu Herrnsheim, called the Dutch fishermen "wholly predatory" and claimed that they were snatching "seven-eighths" of the yearly catch.[37] Statistical data, however, do not support this claim. In fact, the annual German catch exceeded the Dutch catch almost every year between 1875 and 1935. The figures also make plain that delta fishermen felt the impact of stock depletion almost as quickly as their upstream counterparts. After peaking at 100,000 in 1885, the delta catch sank to around 50,000 in 1895, to 25,000 in 1915, and from there it fell into oblivion.[38]

Two major pollution scandals also generated the mistaken belief that phenol was killing off the Rhine's salmon stocks (in fact, phenol passes through the fatty tissue of salmon without killing them). The first scandal erupted in September 1911, when fishermen and restaurant owners complained to the Rhineland Minister-President in Düsseldorf that the Rhine between Kaiserswert and Düsseldorf-Reisholz had become so contaminated with petroleum byproducts that (in the words of one complainant) "the entire fish catch, and especially Rhine salmon, tastes and reeks of phenol and petroleum."[39] An investigation revealed that the phenol was emanating from the Dutch-owned Rhenania Refinery at Reisholz, and the refinery was forced to construct a dephenolization plant on its facilities.[40] A second phenol scandal erupted fourteen years later, in March 1925, when the Cologne firm Rheinlachs-Räucherei Lisner (a salmon and eel smokery) complained to the Rhineland Fisheries Association that salmon caught in the vicinity of the Emscher "tasted so bad that customers were demanding their money back."[41] A new round of investigations revealed that the main polluter was, once again, the Rhenania Refinery at Reisholz. Baffled at first, police resolved the mystery when an employee provided them with confidential information on factory practices. "Each month this factory uses somewhere between fifty and one hundred tons of sulfuric acid in the refining of its benzine products," he stated. "The sulfuric waste is then stored, along with sodium hydroxide, in huge drums. As soon as thirty or forty drums are full, the waste gets dumped into the Rhine under cloak of darkness."[42] Once again, the complaints of fishermen ceased when the factory agreed to stop engaging in this illegal practice.

Complaints of overfishing and phenol pollution merely deflected attention from the primary causes of salmon decline: dredgers and dams. On the Middle and Upper Rhine, it was a combination of gravel removal and hydro-diversion dams that gutted the salmon stocks. In the Koblenz area, for instance, the annual catch dropped as soon as dredgers eliminated the deep spots in the Rhine bed and rectified the Mosel mouth. In the Cologne-Bonn region, the catch plummeted when dredgers removed the gravel beds at the Sieg mouth and elsewhere (see figure 6.6).[43] Unable to halt the destruction of spawning grounds, the Rhineland Fisheries Association sponsored an intensive artificial propagation program on the Mosel, Saar, Sieg, and Agger tributaries. It may have had a short-term impact: the annual Middle Rhine catch held relatively steady between 1898 (7045 fish) and 1908 (6338 fish), not an impressive return on the millions of salmon eggs hatched by the Fisheries Association, but enough to provide a lifeline to the local fisheries at a time of crisis.[44] In the long run, however, nonstop dredging rendered it all but impossible to find adequate nursing grounds anywhere on the river, making pisciculture a Sisyphean labor. The end came in the 1950s with the construction of locks and dams on the Grand Canal d'Alsace and with the canalization of the Mosel.

On the High Rhine, it was mostly dams that doomed the local salmon industry; dredging played only a small role. The combined Swiss-Baden catch averaged 2408 salmon annually between 1885 and 1910, then fell nearly in half to a 1314 annual average between 1912 and 1924.[45] The downturn coincided with the completion of hydroelectric dams at Augst-Wyhlen (1912) and Laufenburg (1914), which had the effect of severing homeward-bound salmon from most of their Alpine spawning sites. Equally revealing, the dams forced fishermen to change their fishing sites. In 1911, 96.5 percent of the entire Swiss-Baden catch was taken upstream from Augst-Wyhlen and Laufenburg. By 1920, 94 percent of the catch was taken below these sites—the new dams clearly functioning as insurmountable obstacles. Fish ladders at the dam sites were but a hoax: most salmon either dashed themselves against the dam walls in a futile effort to get upstream, or they landed in salmon nets below the dams. Either way they failed to spawn.[46] The coup de grâce for Alpine salmon came in 1932 with the construction of the Kembs hydroelectric plant on the Upper Rhine, which made it impossible to reach Switzerland at all (the fish did, however, get a temporary reprieve when Allied bombs destroyed the plant in 1945).[47]

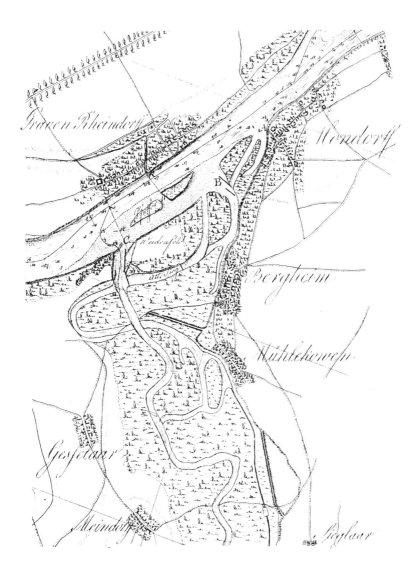

Figure 6.6 The Sieg mouth before rectification. The Sieg's many meanders, loops, and gravel beds once provided an ideal habitat for salmon. It it a centerpiece of the Salmon 2000 project because it still possesses some usable gravel beds. (Source: Wiebeking, *Atlas*, Section XXIII)

As sturgeon, shad, and salmon stocks plummeted, some fisheries turned to eel (not a fish but traditionally treated as one) for their livelihood. Like the other commercial species, the eel is migratory, but its life cycle follows a wholly different pattern. Eel larvae are hatched in the Sargasso Sea, south of Bermuda. Larvae drift with the Gulf Stream for up to three years until they reach the shores of Europe, by which time they have metamorphosed into small eels. From the shores they find their way into inland lakes and rivers, where they spend the next seven to eighteen years living in fresh water before returning to the Sargasso Sea to spawn. The eel escaped the fate of sturgeon, shad, and salmon because it did not depend on the Rhine for propagation, and therefore lost none of its spawning grounds as a result of gravel removal and dam construction. Also, the greater part of its life cycle is stationary rather than migratory. Moreover, when biologists discovered that the delta's weirs and locks were obstructing eel migratory routes, they simply started capturing small eel on the seashores and transporting them to the Rhine artificially—a practical (if not ecologically ideal) solution that has kept the eel fisheries alive to this day.[48]

A mere recitation of destruction and transformation, however, does not fully capture the extent to which two centuries of engineering altered the Rhine and its ecology. The disruptive forces were so great that it is almost absurd to view the Rhine of 1975 as the river it once was. Biologically speaking, the river had become a qualitatively different stream. Once wrapped in trees and other vegetation, its banks were now lined with cement and factories. Once choking with islands, its channel was now choked with heavy metals and urban debris. Once sinewy and crooked, it now looked (as Aldo Leopold once said about all German streams) "straight as a dead snake."

Straightened streams all tend to resemble one another. They lose their individuality and take on the geometric profile prescribed to them by hydraulic engineers: a canal-like look with evenly spaced banks and a trapezoid-shaped (or sometimes U-shaped) channel cross-section (see figure 6.7). The Rhine, Danube, Po, Rhône, and Thames all look pretty much alike today, the only cues to their difference coming from the French, German, Italian, or British cultural trappings on their banks. But more than the loss of a river's "personality" is at stake. Straightened rivers also lose their biodiversity. Since the Renaissance, a major goal of European river engineering has been to channel water into the ocean as quickly as possible. In practice,

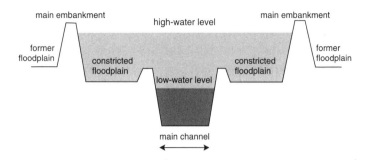

Figure 6.7 Cross-section of an embanked river. Straight lines and geometric shapes replace the irregularities and quirks of a natural river. (Source: Middelkoop and van Haselen, eds., *Twice a River*, 21)

this has meant turning a river into an efficient drainage system by removing its braids, oxbows, islands, rivulets, backwaters, deep spots, shallow pools, and marshlands—in other words, by removing the very sites that provide most organisms their unique niches in the riverine ecosystem. Water pollution imposes a different kind of homogeneity on a river: it reduces or eliminates all organisms that cannot tolerate the contaminants and favors only the few that can.

River organisms depend on a diversified, not a uniform channel. Well-adapted to the natural rhythms of a river, they can withstand floods, droughts, and innumerable other changes, but not a geometric assault from river engineers. As late as 1900, biologists could still distinguish seven bioregions on the Rhine: the Trout Region (the Vorderrhein and Hinterrhein), the Trout/Grayling Region (Alpenrhein), the Grayling Region (the upper half of the High Rhine), the Grayling/Barbel Region (lower half of the High Rhine and upper half of the Upper Rhine), the Barbel Region (lower half of the Upper Rhine and most of the Middle Rhine), the Bream Region (most of the Lower and Delta Rhine from Bonn to Dordrecht), and the Sea-Bream/Flounder Region (near the delta mouth). Nonstop river engineering, however, annihilated these distinctions. The high Alpine and lower delta stretches still have the same predominant species as before. But the entire navigable stretch of the Rhine, from Basel to Rotterdam, now forms but a single bio-community.[49] Small wonder that just nine species—the roach, bleak, common bream, eel, dace, perch, chub, white bream, and gudgeon—account for nearly 97 percent of all fish found there.

The biological degradation took place at a number of different levels. First and foremost, there was a precipitous drop in living space on the river. Even the most adaptable bird and insect species found themselves restricted to smaller and smaller river stretches for the simple reason that most of the Rhine's feeding, breeding, and nesting grounds were gobbled up by land-hungry humans or otherwise destroyed. River straightening took away more than a hundred kilometers of bed and bank space in the main channel alone. Many more kilometers of bed and bank were destroyed when the braids and arms were removed. Dredging removed three thousand river islands, each representing a microenvironment for various organisms. Land reclamation, meanwhile, ate up nine-tenths of the river's original floodplain. All of this added up to an enormous loss of riparian habitat—enough to push many river organisms to the brink of extinction.

Second, today's Rhine consists of a narrow and uniform habitat rather than a broad and diverse one with multiple ecological niches of all shapes and varieties. It therefore supports certain types of species in vast quantities—notably roach fish and zebra mussels—while marginalizing many others. The omnivorous roach will eat zooplankton, larvae, insects, mollusks, or plants as the occasion dictates. Females will lay up to 100,000 eggs at a time—on bankside roots, on aquatic plants, on stony surfaces, or wherever they must. The zebra mussel is a similarly indiscriminate. In 1983, Mainz researchers recorded 20 million verliger (planktonic larvae) per second flowing in the Rhine's water during the annual spatting season, or seventeen billion verliger larvae per day.[50]

Third, the river's dams and barrages have cut the links between upstream and downstream habitats, making it all but impossible for fish and other aquatic species to reach all parts of the river to which they are adapted. The disruption was fatal for most migratory fish, since their life cycles require that they move freely up and down the river from headwaters to the sea. But stationary fish also now find themselves locked into a truncated habitat artificially set for them by human engineering. Fourth, the line that once separated the Rhine's freshwater species from its estuarian species has been breached. The river's chloride content has become so high that its upstream stretches have taken on some of the brackishness once characteristic only of the delta mouth. Finally, the line that once separated lithophiles from soft-bed dwellers has disappeared. Cement-and-stone embankments now function as the universal highway for the lithophiles.

Riparian-zone destruction, water eutrophication, the arrival of non-native species—these are the classic symptoms of a biologically disturbed stream.[51] River engineering has severely compromised the complex web of life that once prevailed on the Rhine. Nothing was left unaffected by the engineering work, not even the heartiest of the river's inhabitants: fungi, bacteria, algae, protozoa, flatworms, nonsegmented and segmented round-worms, and wheel animals. The near-disappearance of riparian vegetation has meant that there are far fewer leaves and roots for fungi to decompose, which in turn has meant that there is less organic matter for other micro-organisms (shredders and collectors) to ingest, which in turn has meant less food available to invertebrate macrofauna. Water pollution (especially eutrophication) has placed a similar limit on the river's food supply by killing off many of the caddisfly, stonefly, and mayfly taxa upon which so many other river organisms once depended. These disruptions, in turn, have reduced the number of fish, birds, amphibians, and mammals that the river can support.

The Rhine's degradation as a biological habitat was aggravated by several interrelated processes, among them a decline in water quality from the addition of chemicals, salts, and other industrial and urban pollutants; stream eutrophication due to fertilizer runoff from agricultural fields; an increase in water temperature from the use of river water in the cooling towers of conventional and nuclear power plants; and the near-disappearance of the river's bacteria-rich wetlands (its "self-cleansing" sites). But behind all of this devastation was the driving need to harness the river entirely to human purposes, and the willingness to accept water pollution and flood-plain destruction as an unavoidable consequence of industrial and agricultural development. Aiding and abetting this destruction was the ideology of hydraulic engineering. As long as Europeans continued to perceive the Rhine as an imperfect canal, as long as they manipulated its channel as if it could be separated from its floodplain, and as long they treated the river as a sewage system rather than as a biological habitat, there could be no end to the spiral of devastation.

Mary Shelley's *Frankenstein* was a science fiction novel about the horrors of technological hubris. It seems appropriate that she conceived her story while on a visit to the Swiss Alps, that she chose a fictional German scientist as her main character, and that she published her book in 1818 just as the rectification work was getting under way. But the parallels end there, for Frankenstein's hubris lay in his attempt to bring dead matter back to life, and not the other way around.

I refuse to sing
another false paean of praise for you....
Mine is an obituary for the fish
that have perished on the poisons
that pour into your channel
from sewers aplenty, left and right....
Not even the ocean welcomes your arrival.

— Willy Bartock (1963)

CHAPTER 7

A River Restored?

The towering figure of twentieth-century Rhine biology was the great naturalist Robert Lauterborn. Born in Ludwigshafen in 1869, he grew up in the shadow of the gargantuan BASF chemical factory, his hometown's most important industry and the Upper Rhine's most notorious polluter. After completing his doctoral dissertation on the topic of *Ceratium hirundinella* protozoa at Heidelberg University in 1896, Lauterborn spent the next several years investigating the quality of Rhine water for the German states of Baden, Hesse, and the Bavarian Palatinate (now part of Rhineland-Palatinate). From 1908 to 1945, he served as head of the Palatinate's water research team. At the time of his death in 1952, at the age of eighty-two, he had published over a hundred articles on Rhine flora and fauna, and participated in more Rhine water analyses than almost anyone of his generation. Though much of his research has been superseded by that of Rainer Kinzelbach, Anton Lelek, Thomas Tittizer, and many others, Lauterborn left behind an indispensable "snapshot" of the Rhine's plant and animal life at the turn of the century, and his numerous forays into natural history helped others to reconstruct a relatively reliable record of the river's native flora and fauna. So much did he dominate his generation that a biology journal bears his name: *Lauterbornia*.[1]

No radical anti-industrialist, Lauterborn recognized that some of the Rhine's floodplain had to be sacrificed to economic development, especially in urban areas. As a trained biologist, however, he also understood that a floodplain plays a crucial role in maintaining the health and stability of a river, and throughout his life he strove to keep as much as possible of the Rhine's original habitat intact. "Shielding our water resources from 'the wastes of civilization' is one of the most important aspects of nature protection," he wrote in his memoirs:

This became crystal clear to me early in life as I saw the rate at which all traces of a natural environment were disappearing from the vicinity of my hometown as it expanded and grew. Back in those days, one could still find a sufficient number of areas on the outskirts of cities that had not yet been denatured. These regions functioned as preserves in which the whole array of local flora and fauna could be found. It seemed to me that one of the most pressing tasks of any natural scientist was to protect those last remaining open spaces so that succeeding generations would have a chance to experience and enjoy them.[2]

As the Rhine's most celebrated biologist, Lauterborn was often called as a government and independent expert to assess the environmental impact of various development schemes. But as he learned from bitter experience, it was far easier to preserve the Rhine's flora and fauna in photographs and journals than in reality. A particularly nasty dispute erupted in 1930 over plans to drain a marshy area at Neuhofen Altrhein, an old river branch that was at the time the largest stretch of undeveloped riverscape remaining in the Ludwigshafen-Mannheim vicinity. The request to drain the marshland came from the personnel at the Water Resources Office of Neustadt, whose task it was to promote economic development in the region; and from the mayor of Altrip, who was also the director of the region's largest brick factory and gravel-removal business and who therefore stood most to gain from the drainage scheme. Because the project involved turning a substantial amount of reclaimed land over to brick production, and because the drainage would cause an estimated 1.2 meter drop in the surrounding water table, state and national authorities became involved. As representative of the Reich Technical Bureau for Nature Protection, so did Lauterborn.

The outcome (as was so often the case) was largely predetermined: national and regional authorities were predisposed to turning the marsh over to the brick factories unless a compelling counterargument could be made on purely economic grounds. After much wrangling and debate, they decided to go ahead. It mattered little that there were sound ecological reasons for not developing the Neuhofen Altrhein, or that the claims of the prodevelopment forces were specious. The Altrip mayor, for instance, argued that the region was suffering from an acute shortage of arable land. Lauterborn and others pointed out that Altrip had been a thriving agricultural and fishing village *until* the brick industry had settled in the region and bought up most of the farmland for use as loam and gravel pits. Altrip's farmers once had 423 hectares under cultivation, but by 1930 only 150 hectares remained in their hands. The rest had been bought by the region's two largest brick and gravel factories. The new marsh-drainage plan, moreover, made plain that any newly reclaimed land was earmarked for factories, not farms.

Unable to make headway with the land-shortage ploy, the Water Resources Office played the public hygiene card: the marshes were breeding grounds for waterborne diseases. Altrip, so the claim went, was in the grips of a "mosquito plague," putting people who used the Rhine for recreational purposes at risk. But Lauterborn and others pointed out that local inhabitants hiked and played in the region every day, much as they had in the past, without any discernible danger to their health. On-site investigations, moreover, revealed that the local mosquito populations utilized the stagnant water in the abandoned loam pits, not the slow-flowing water in the marshes, as their breeding sites. But economics invariably trumped ecology: the authorities gave the go-ahead to marsh drainage, and then turned the reclaimed land over to the brick factories—all without giving much thought to the fact that they had just expunged one of the few remaining open spaces in the region.[3]

Neuhofen Altrhein was only one of many battles that Lauterborn fought and lost, and he was only one among many Rhine biologists (albeit the most prominent one) engaged in rear-guard actions to save the Rhine's remnant floodplain. No doubt the Great Depression, which struck Germany during the controversy, loomed large in the minds of government leaders as they deliberated. But their decision in favor of the brick factories was anything but exceptional. A laissez-faire attitude toward environmental matters was the rule: few and far between were the development plans that did not receive approval in one form or another.

At the root of the problem, from an environmental perspective, was the river regime established by the Congress of Vienna in 1815: the diplomats had created an overarching blueprint for improving the Rhine as a navigational and commercial artery, but no corresponding one to protect it as a biological habitat. The river's biological viability was thus destroyed as much by indifference as by design—by a succession of incremental engineering decisions that cost the river a few hectares here and a few hectares there until almost all of the floodplain had disappeared. The logic of navigation and commerce pushed in only one direction: maximum use of the river for human production and consumption regardless of the long-term consequences for riverine ecology.

After 1945, however, a new mentality toward the Rhine began to emerge. This shift was not primarily a change of heart about the protecting riparian plants and animals, though concern over the disappearance of natural habitat did play an increasingly important role after the 1960s. Anthropocentric concerns were largely behind the change in attitude. Industrial and urban leaders were coming to realize that they could not continue to dump untreated effluents into the river and still expect it to provide their freshwater needs. Tourists and townspeople alike increasingly grumbled that the Rhine looked more like a sewer than a Romantic icon. And governments finally began to see that incessant hydraulic tinkering was not the panacea it was once thought to be. The river was simply no longer capable of fulfilling the multiple roles assigned to it by the vast industrial-urban-transportation-tourist complex on its banks.

As more and more companies built their factories on the Rhine's banks, and as riparian cities expanded in size and number, water quality plummeted, and by the 1970s the liquid flowing in the riverbed was nothing more than a toxic soup of chemicals and raw sewage. Organic compounds and heavy metals were found everywhere in the Rhine basin—in the channel, in the silt, and in the surrounding landscape. Bioaccumulation of chemicals and heavy metals in the food chain had become so severe that many organisms could no longer survive in the river system; and the buildup of organic nutrients had reached such levels in certain stretches (notably the Lower Rhine) that fish and other higher organisms were scarce. The same destructive processes that had previously undermined the self-cleansing capacity of so many Rhine tributaries—the Emscher, Erft, Wupper, Sieg, Main, Mosel, and Neckar, among them—were inexorably beginning to threaten the trunk system itself. But now that the entire basin was dying, the riparian states could no longer entertain the false hope of exporting the problem farther downstream.

Yet another problem had to be confronted at long last: flooding on the Middle, Lower, and Delta Rhine. The flood danger on the lower stretches increased each year as the Upper Rhine's natural floodplain disappeared and as industries and cities put themselves in harm's way by settling directly on the riverbanks. Without a concerted effort to restore some of the lost floodplain (or at least create impoundment zones to store water during flood conditions), the densely populated regions of Hesse, North Rhine–West-phalia, and the Netherlands were destined to experience inundations of greater and greater magnitude. And farmers in these regions would have to reckon periodically with a layer of flood silt so full of heavy metals and organic compounds that their fields would become too contaminated to grow crops.

Political relations among the western European states made the post-1945 period favorable for river restoration, much as the post-Napoleonic era had created favorable conditions for river development a century and a half ear-lier. World War II brought an end to the bitter military struggles between Germany and its neighbors. As the riparian states began to work together within the framework of NATO, the European Community, and (later) the European Union, they sought new avenues for resolving common environ-mental problems as well. Once peace returned to the Rhine, moreover, the chemical industry lost its special status as a civilian branch of the mili-tary; its political power waning, it could no longer flout pollution laws with impunity. The post-1945 period also brought an end to the coal age. Ruhr production fell as petroleum imports climbed, and by 1996 coal output stood at less than half its pre-1945 levels. Water quality on the Lower Rhine would have improved significantly even if no other cleanup efforts had been undertaken. Finally, the 1960s witnessed the birth of a powerful environ-mental movement in Europe: the same activists who challenged the nuclear industry also challenged the notion that unrestrained river development was always a sign of progress. Although cleanup of the Rhine began well before any Green parties established themselves in the riparian states, the gradual "greening" of Western Europe gave an enormous boost to those efforts—especially after the Sandoz spill revealed just how precarious ripar-ian life on the Rhine had become.

Improvements in Water Quality

The first sign that a new spirit of cooperation was emerging on the Rhine came with the establishment of the International Commission for the Protection of the Rhine against Pollution (the Rhine Protection Commis-sion) in 1950. The Netherlands government—as the primary recipient of upstream pollution and floodwater—was the driving force behind this new organization. France, Germany, Luxembourg, and Switzerland joined the Netherlands as charter members, and the European Community joined as a contracting party in 1976. The Rhine Protection Commission kept a low profile in its early years, in part because its initial mission was restricted to a scientific assessment of the river's water quality, and in part because the riparian governments were too preoccupied with postwar reconstruction to devote much time to environmental issues. Then in 1963 the member states signed the Bern Convention on the International Commission for the Protection of the Rhine against Pollution. The Bern Convention gave the commission the authority to hold annual plenary sessions and draft international treaties. In 1972, the commission was given the added task of organizing regular ministerial-level meetings. These Rhine Ministers' Conferences remain the single most important forum for handling issues of Rhine pollution and ecology.[4]

Several other new Rhine organizations—all affiliated in some way with the Rhine Protection Commission—also came into existence after World War II. Among the most important is the International Working Group of the Waterworks of the Rhine Basin (usually known by its German acronym, IAWR). Founded in Düsseldorf in 1970 along the lines of a traditional Rhineland-Westphalian riparian cooperative, it quickly grew into an international consortium of more than a hundred Rhine waterworks in the Netherlands, Belgium, France, Germany, Austria, and Switzerland. A second riparian cooperative—the Association of Rhine and Meuse Water Supply Companies (Dutch acronym: RIWA)—was created to oversee water quality on all streams that flow through Belgium and the Netherlands on their way into the common Rhine-Meuse delta. Meanwhile, a third riparian cooperative, the Working Group of the German Federal States (German acronym: LAWA), was established to handle common water issues among the German states. The riparian cooperatives have never been as much in the public eye as the Rhine Protection Commission, but they have been instrumental in improving industrial and urban water supplies, and in reducing the river's overall pollution load. Two other organizations also deserve mention: the International Commission for the Protection of Lake Constance, established by Baden-Württemberg, Bavaria, Austria, and Switzerland in 1960 to reduce pollution levels in Lake Constance; and the OSPAR Commission (short for Oslo and Paris Treaties), established by the North Sea countries to monitor river-based and marine-based pollution in the northeast Atlantic.

Between 1972 and 1976, the Rhine Ministers drafted three international treaties based on the scientific data and policy recommendations provided by the Rhine Protection Commission and its affiliates. The first treaty, the

1976 Bonn Convention Concerning the Protection of the Rhine against Pollution by Chlorides, focused on waste salts from industrial production (mostly potash fertilizer). The second, the 1976 Bonn Convention for the Protection of the Rhine against Chemical Pollution, addressed all chemical inputs into the river, both those from "point sources" (chemical and other manufacturing sites) and those from "nonpoint sources" (such as agricultural and road runoff). The third, the 1976 Bonn Convention for the Protection of the Rhine against Thermal Pollution, was drafted in anticipation of a proliferation of nuclear power plants; after the nuclear industry stalled in the late 1980s, it was dropped in favor of a plant-by-plant approach to water temperature control. Taken together, these diplomatic initiatives signaled the most extensive revision of the Rhine river regime since the Congress of Vienna. For the first time in more than a century and a half, the riparian states began to "see" the river not simply as a navigational and commercial highway but also as a biological habitat in need of protection.

The Chloride Convention took a pragmatic approach to waste salt management. It did not obligate Rhine industries to eliminate all influx sites (indeed, it ignored all but the largest chloride producers). It simply required the riparian states to reduce their collective discharges until the river's overall load fell below the point where it threatened the survival of freshwater species. In the absence of any clear scientific data as to where that threshold lay, the Rhine Ministers set the upper limit somewhat arbitrarily at 200 kilograms per second (k/s), using the river's salt load of a century earlier as a rough guide. Since the salt load stood at 365 k/s (using 1970 figures), a reduction of 165 k/s was required. The bulk of the cleanup burden logically fell on the Alsatian potash industry, which generated the lion's share (130 k/s in 1970) of the river's human-induced waste salts. The convention obligated France to construct chloride-removal systems at their potash plants, and to pump the recovered salts into underground limestone formations near Mulhouse. The other signatories—the Netherlands, Luxembourg, Germany, and Switzerland—were given the much easier task of modestly reducing chloride discharges on their territories (though they did agree to share the costs of salt removal in Alsace).[5]

Moving from paper to practice proved more difficult than anticipated, largely because Alsatian farmers opposed underground storage for fear that the salts would eventually leach into the local groundwater. Ever sensitive to its farming constituency, the French government floated several alternative disposal plans, most ingeniously "Project Eau claire," which called for the construction of a pipeline running down the Rhine from Alsace to the North Sea. Such a pipeline would deliver the salts directly to the ocean, where they would be harmless to marine life (and Alsatian farmers). Unfortunately, a pipeline stretching from Strasbourg to Rotterdam made economic sense only if utilized for disposal of other industrial pollutants as well as salts—and the North Sea states (organized in OSPAR) strongly opposed any disposal method that further burdened the northeast Atlantic. In 1985, therefore, the French abandoned Project Eau claire and ratified the Chloride Convention, after first finding methods for storing the waste salts more securely.[6] Since then the river's salt load has dropped significantly (see figure 7.1).

The 1976 Chemicals Convention—the other major convention that emerged out of the Rhine Ministers' conferences—obligated the riparian states to "eliminate pollution of the surface waters of the Rhine basin by dangerous substances," defined as all substances that endanger human health by virtue of their "toxicity, persistence and bioaccumulation." The most important of these substances were mercury, cadmium, petroleum-based mineral oils and hydrocarbons, and organohalogen, organophosphorus, and organotin compounds. The Rhine states also promised to "reduce" (though not eliminate) all substances that had a "deleterious effect on the

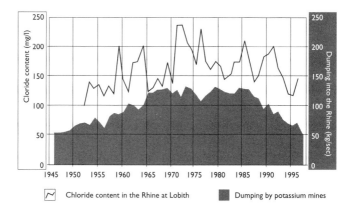

Figure 7.1 The relationship between potash mining and the Rhine's chloride content. Potash fertilizer production has accounted for much of the Rhine's salt load for the past century. (Source: Middelkoop and van Haselen, eds., *Twice a River*, 71)

aquatic environment" if their impact could be "confined to a given area." These substances included zinc, copper, nickel, chromium, lead, selenium, arsenic, antimony, and a dozen other metalloids and metals. A wide variety of other chemicals was also covered, including biocides and their derivatives; toxic and persistent organic compounds of silicon; inorganic compounds of phosphorus and elemental phosphorus; nonpersistent mineral oils and hydrocarbons; cynanides and fluorides; and ammonia, nitrites, and other substances that have an adverse effect on the dissolved oxygen levels. The treaty was comprehensive in scope, covering both "point source" pollutants emanating from specific industrial and municipal sites, and "diffuse source" pollutants from agricultural runoff, streets and highways, surface water, and atmospheric precipitation.[7]

Three protocols—for cadmium (1983), mercury (1985), and carbon tetrachloride (1986)—were subsequently added to the Chemicals Convention. The protocols were needed because each chemical presented unique problems of detection and abatement. Cadmium, for instance, enters the soil, water, and air of a river basin through zinc refining, coal and oil combustion, iron and steel manufacturing, and phosphate-based fertilizing. The production of nickel-cadmium batteries and plastic stabilizers also adds to cadmium levels, as do electroplating, copper refining, coke production, cement manufacturing, and waste incineration. The cadmium protocol therefore spelled out emission limits in each sector, established measurement methods, and listed acceptable mitigation technologies. The protocols governing mercury and carbon tetrachloride were similarly stringent.[8]

Ratified in 1979, the Chemicals Convention was initially on a faster track than the Chloride Convention. But it, too, ran into implementation snags. One cause for delay was the absence of technologies for reducing or eliminating many of the chemicals and chemical classes enumerated in the convention. Pulp-and-paper mills, for instance, could not reduce their output of organic halogenated compounds until they first found a chlorine substitute for bleaching. Other companies faced similar problems with the substances they used in production. Treatment plants, moreover, often took years to design and construct, especially if the mitigation technologies were new or untested. Space for constructing new facilities was also often at a premium, especially in highly industrialized areas, where the need for the plants was often greatest but unused land in shortest supply. Foot-dragging also played a significant role: implementation relied heavily on the goodwill of the Rhine riparian governments, all of which were susceptible to anti-environmental lobbying from the polluting industries. Progress was so slow that in 1987 (shortly after the Sandoz spill), the European Parliament publicly berated the riparian states for having accomplished "scandalously little as regards both the prevention of disasters and accidents and the reduction of chemical, salt and thermal pollution," and demanded a quicker pace "given that the pollution of the waters of the Rhine is already too far advanced and that, in many places, the river bed has already suffered damage which will prove difficult to repair."[9]

After 1986, the Rhine Ministers and the Rhine Protection Commission redoubled their cleanup efforts, the principal result of which was the Rhine Action Plan for Ecological Rehabilitation (1987). The Rhine Action Plan put many of the chemicals covered in the Chemicals Convention (including all the insecticides and heavy metals) on a fast track for reduction. During Phase I (1987–89) national and regional authorities were to make sure that mitigation technologies were put in place at factories and treatment plants. During Phase II (1989–95), they were to ensure that discharges of these substances were reduced by 50 percent or more. The Rhine Action Plan was more aggressive than the Chemicals Convention in another way as well:

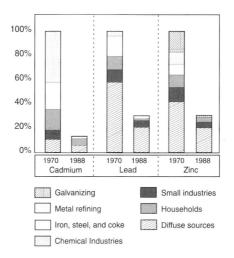

Figure 7.2 Decreases in aqueous emissions of three heavy metals in the Rhine, 1970–88. The 1976 Chemicals Convention has proved phenomenally successful at reducing these and many other dangerous substances. (Source: Anderberg et al., *Old Sins*, 63)

it targeted every factory on the Rhine, regardless of size, that produced or consumed any testable amount of organic and inorganic chemicals, petrochemicals, mineral oils, cellulose, paper or pulp, metals, paints and varnishes, dyes, textiles, leather, coal or coal tar, rubbers and plastics, chemical cleaners, and glass products.[10] In 1991, several additional substances were added to the priority list—including arsenic, azinphos-ethyl, fenitrothion, malathion, and dioxins—in order to bring Rhine regulations in conformity with OSPAR's North Sea regulations.[11]

Improvements in water quality between 1970 and 2000 demonstrate unequivocally that both the Chemicals Convention and the Rhine Action Plan have had an enormously positive impact on the entire Rhine basin (see figure 7.2). On the heavily polluted Lower Rhine, for instance, the average yearly concentration of ammonium (measured at Bimmen-Lobith) dropped from just over 2.5 milligrams per liter (mg/l) in 1972 to 0.5 mg/l in 1986, a fivefold decrease. During the same period, mercury concentrations decreased more than tenfold and cadmium concentrations nearly eightfold. The content of lead, copper, zinc, chromium, nickel, and arsenic in water sediment also dropped sharply—enough to make it possible for the first time in decades to resume the practice of using Rhine sediment and silt for land reclamation in the Netherlands. Most dramatically, the water's oxygen content, after reaching an all-time low of around 4 mg/l in 1971, climbed to 9.5 mg/l in 1986, close to the level of saturation that the Lower Rhine would have under natural conditions.[12]

Similar improvements occurred elsewhere on the Rhine. The Rhine Action Plan set 1995 as the date for reducing the targeted chemicals by 50 percent. In actuality, that reduction level was achieved two years ahead of schedule for all but three of the chemicals (ammonium, endosulfan, and 4-chlorotoluene). In about half the cases, a reduction level of 80 percent or more was attained. The mean daily load of organic halogenated compounds, for instance, dropped by 82 percent between 1986 and 1993, almost entirely because pulp-and-paper mills (the principal producers of these waste compounds) found an alternative to chlorine for paper bleaching in the mid-1980s. Similarly, phosphorus levels dropped when researchers found substitutes for phosphatic detergents (in widespread use from the 1960s to the 1980s) and developed new techniques for removing phosphorous from drinking supplies.[13]

One group of substances, however, has continued to bedevil the Rhine: nitrates. An organic pollutant, nitrates tend to accumulate in slow-moving stretches and river widenings—primarily Lake Constance, the Lower Rhine, the IJsselmeer, and the North Sea estuaries—where they supply algae with food and thus contribute to eutrophication. Runoff from fertilized agricultural fields, a notoriously difficult source to control, is the principal cause of nitrate contamination. The riparian states have mitigated the problem somewhat by taking some land out of cultivation, by restricting the use of artificial fertilizers on fields near the river, and by capturing and treating as much of the runoff as possible before it enters the river. But these methods are only partly effective, and as a result contamination levels have not dropped nearly as much as those of other pollutants.[14]

Biological analysis (the other common way of measuring water quality) also clearly shows that the chloride and chemical conventions have significantly improved the Rhine's condition. At its worst, in 1976, most of the Upper and Middle Rhine belonged to Class II/III ("moderately" to "strongly" polluted), while the Lower Rhine stood at III/IV ("strongly" to "completely" polluted). Water quality on the entire navigable Rhine was thus compromised to a considerable degree. By 1990, the Upper and Middle stretches had moved up to Class II, a clear sign that dissolved oxygen levels

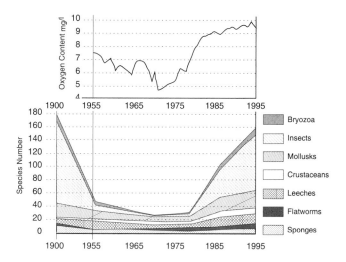

Figure 7.3 The relationship between the Rhine's oxygen content and the number of invertebrate species the river supports, 1900 to 1995. Pollution-sensitive insect species have benefited the most from cleanup efforts. (Source: ICPR, *The Rhine: A River and Its Relations*, 21)

had returned to near normal. The Lower Rhine also moved up an entire step to Class II/III for much the same reason. Meanwhile, the Main tributary went up a step and a half, from IV to II/III. The Neckar and Wupper each moved up a full step, from III/IV to II/III and IV to III respectively. The Lippe moved up a half step, from III to II/III, as did the Erft, from IV to III/IV. Only the Emscher showed no improvement: it was still at Class IV in 1990.[15]

Improvements in water quality, modest as they were in some Rhine stretches and tributaries, brought immediate benefits to the river's benthic (bottom-dwelling) communities. The number of invertebrate macro-fauna—mollusks, insects, crustaceans, sponges—had fallen from 165 species in 1900 to a mere 27 in 1971. By 1978 the number had climbed back slightly to 34, by 1986 to 97, and by 1989 to 155, the upward surge corresponding closely to the scope and pace of the cleanup efforts (see figure 7.3).[16] The river, of course, has yet to return to its prior level of ecological health and stability. During the 1990s, for instance, Dutch researchers noted a population explosion of *Corophium curvispinum*, a tiny filter-feeding amphipod that thrives on phytoplankton and suspended organic matter in highly eutrophic water. It apparently hitchhiked from the Caspian Sea to the stone-lined artificial substrata of the Middle Rhine in 1987, then spread quickly to the Lower and Delta Rhine, attaining within a five-year span population densities as high as three-quarters of a million individuals per square meter—enough to make it one of the most numerous invertebrates in the Rhine.[17] Researchers on the Middle Rhine have also noted that many sensitive species have yet to mount a comeback, that certain fish and mussel populations continue to show evidence of DNA damage, that some chironomid larvae and sponges are still malformed, and that bank-filtrated drinking water still has detectable amounts of persistent compounds.[18] In general, however, the upward trend in biodiversity reflects the fact that the water has improved greatly over the past twenty-five years, and that the river is slowly returning to something like its natural chemical condition.

Habitat Restoration

The most highly publicized aspect of river restoration to date has been Salmon 2000, a campaign to establish self-sustaining populations of Rhine salmon by the first decade of the new millennium. Salmon 2000 appeared in embryonic form in the Rhine Action Plan ("the ecosystem of the Rhine must become a suitable habitat to allow the return to this great European

river of the higher species which were once present here and have since disappeared, such as salmon"), but it was not until the Rhine Protection Commission issued its blueprint for riparian restoration, the Ecological Master
Plan for the Rhine (1989), that salmon repopulation commenced.[19]

Salmon 2000 targeted all of the river's main migratory fish (salmon,
sea trout, shad, sea lamprey, and sturgeon) for repopulation, but the spot-
light from the outset has been on salmon. Salmon once inhabited nearly
every niche of the Rhine and are therefore a key indicator of the river's
health; as they recover, so too will other migratory species. They also have
a greater "charismatic" (symbolic) value than other Rhine fish: many Europeans equate river restoration with a return of the salmon. Though there
were many reasons for the disappearance of the Rhine stocks (including
overfishing), two were particularly lethal: dredging, which deprived the
salmon of their best spawning and smolting grounds; and dams, which
chopped the river into segments and thus severed the migration routes.
Salmon 2000 has therefore consisted mostly of three interrelated activities:
identifying and protecting those Rhine tributaries that still possess a sufficient amount of well-oxygenated gravel beds; constructing fish passages
around dams and other in-river obstacles (or removing the obstructions
altogether); and restocking the river with eggs and alevins from nearby
salmon rivers (mostly those in Brittany, Scotland, Ireland, Norway, and
southwest France). Fish ladders are the most visible (and costly) feature of
Salmon 2000. Typically they consist of a row of adjacent basins, each of
a slightly different elevation. The basins function much like a small river
braid, breaking the ascent and descent around the dams into manageable
steps.

Although the Rhine Protection Commission coordinates the Salmon
2000 project "from above," the restoration work itself has been left to
national and state authorities. By necessity, most of the initial work took
place in the Netherlands. The Dutch project consisted mostly of removing
hindrances at the Rhine's main mouths—the Waal, Nederrijn-Lek, IJssel,
Meuse—the entry and exit points for all fish migrating between the river
and the North Sea. Coastal sluiceways on the IJsselmeer (especially at the
Afsluitdijk) were reconstructed to permit fish migrations up the IJssel. Fish
passages were also added to the weirs on the Nederrijn-Lek and Meuse
rivers. Most notable was the redesign of the Haringvlietdam to allow migrations up the common Meuse-Rhine delta.[20] Once these alterations were
completed, work began farther upstream, first in North Rhine–Westphalia,

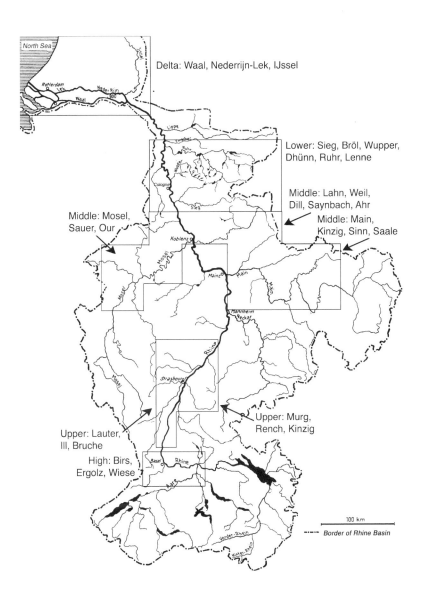

Figure 7.4 Major restoration sites of the Salmon 2000 project. Only a few Rhine tributaries are suitable for salmon today. (Adapted from ICPR, *Ist der Rhein wieder ein Fluß für Lachs?* 7)

then in Rhineland-Palatinate, Hesse, Luxembourg, France, Baden-Würt-temberg, and Switzerland (see figure 7.4).

Salmon restoration in North Rhine–Westphalia has occurred mostly on the Sieg and one of its feeder streams, the Bröl. The Sieg basin was singled out among the Lower Rhine's many tributaries because it still possesses an unusually large number of suitable gravel beds—about 14.2 hectares (35 acres) of spawning grounds and 46 hectares (114 acres) of smolting grounds, enough to sustain 46,000 or more juvenile salmon each year. The project's first phase consisted of constructing thirteen fish passages to circumvent the Sieg's many weirs. Salmon planting began a short time later, and by the early 1990s a modest annual salmon run, estimated at fifty to a hundred individuals, had been recorded, with runs of five hundred expected in the future.[21] Success on the Sieg prompted North Rhine–Westphalia to restore a few of its other tributaries as well. Salmon and sea trout were recently introduced on the Wupper and one of its feeder streams, the Dhünn, in the hope of attaining a self-sustaining population there by 2010. The Ruhr too has been targeted for repopulation, with plans now under way to add fish passages to its weirs and locks.[22]

In 1992, Rhineland-Palatinate and Hesse undertook similar salmon-res-toration projects on the Middle Rhine, especially on the Saynbach and Lahn tributaries. The removal or reequipping of twenty-seven weirs on the Saynbach opened up 12.6 hectares for use as spawning and nursery bio-topes for salmon and sea trout. Ten weirs on the Lahn were similarly recon-structed, opening up 13.6 hectares. A few Lahn tributaries, notably the Dill and Weil, may eventually provide a small additional amount of habitat. Two returning salmon were spotted on the Lahn in September 1997, sug-gesting that a small self-sustaining population is taking root.[23]

The Hessian government (with some assistance from Bavaria) also initi-ated a small restoration project in the Main basin in the 1990s. The Main lost its capacity to spawn salmon and sea trout when it was outfitted with thirty-four navigation locks (the last of which came in 1962). Three of its feeder streams—the Kinzig, Sinn, and Saale—would still support salmon except that six lock-and-dam systems stand in the way. The lower four (at Kostheim, Eddersheim, Griesheim, and Offenbach) were outfitted with fish passages when they were built, but they were poorly designed and have never worked well. The upper two (at Mühlheim and Groß-Krotzenburg) have no fish passages at all. In 1998, a research team tracked the progress of approximately 14,000 migratory individuals from the Rhine up the Main.

They discovered that only a minuscule number of fish were able to wiggle through the hurdles. Even more revealing, of the twenty-five species that were tracked, most failed to surmount even the first hurdle. Eel, roach, and asp were the only ones that managed to get through with any regularity, and their numbers diminished significantly with each in-river hurdle. Hesse plans to outfit all six dam-lock systems with new fish ladders. Once the longitudinal pathways have been opened, salmon repopulation will begin.[24]

Luxembourg undertook a similar restoration project on the Sauer and Our, two small tributaries on the upper reaches of the Mosel, in the early 1990s. Once a favorite destination for migratory fish, the Mosel (like the Main) lost its salmon and sea trout stocks in the 1950s when it was outfitted with twelve hydrodam-lock systems for barge traffic. The Mosel banks themselves, now almost wholly canalized, are no longer suited for salmon, but the Sauer still possesses about five hectares of spawning and sixty hectares of nursing habitat, the Our one hectare of spawning and eleven hectares of nursing habitat. As an experiment, the Luxembourg government released 200,000 young salmon into the two streams between 1992 and 1999. To date not a single one has managed to find its way back to spawn—an unequivocal indication that the Mosel's in-river barriers present an insurmountable hurdle for salmon. The migration routes, however, are being restored: in 1999, Germany, France, and Luxembourg decided to redesign the Mosel's channel to accommodate two-way traffic over the next twenty-five years, and to equip the new dam-lock systems with fish passages.[25]

Restoration work on the Upper Rhine began in 1997, when France and Baden-Württemberg authorized the construction of fish ladders at the Iffezheim and Gambsheim dams. The Iffezheim passage, now completed, consists of thirty-seven basins that circumvent the main channel (see figure 7.5). The Gambsheim passage will become functional around 2003. Together they will extend the Rhine's longitudinal pathway to Strasbourg—far enough up the river to open up several tributaries in Alsace-Lorraine and Baden-Württemberg for salmon and sea trout repopulation. The French and German authorities also plan to retrofit the Upper Rhine's next four hurdles—the Loop Diversions at Strasbourg, Gerstheim, Rhinau, and Markolsheim—with fish passages. Thereafter it may be necessary to redesign the Grand Canal d'Alsace's four hydrodams (Vogelgrün, Fessenheim, Ottmarsheim, and Kembs), but for now they hope fish will be able to bypass the dams by using the remnant bed and the existing fish passage

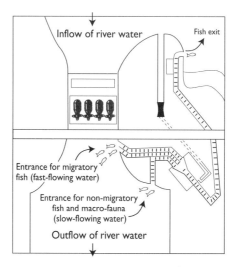

Figure 7.5 Diagram of the new fish passage on the Iffezheim hydrodam (Adapted from *Topic Rhine*, June 1997)

at Kembs. The French have released some salmon into the remnant bed to
determine whether it is suitable for spawning and nursing.[26]

The Association Saumon-Rhin, a private nonprofit organization affiliated with the Rhine Protection Commission, has overseen much of the restoration work in Alsace-Lorraine. It has concentrated its attention on the
Ill tributary and its feeder streams, especially the Bruche, Gießen, Fecht,
Doller, and Blind. As elsewhere, the association began by identifying the
extant spawning spaces and in-river barriers. But it has also experimented
extensively with the restoration of meanders, deep-water pools, gravel pits,
and secondary channels in an innovative attempt to augment the amount
of living space available to migratory fish. An initial repopulation program,
begun in 1991, met with only a modest success. Now that Iffezheim no
longer blocks the route to the Ill, the prospects for establishing a self-sustaining population are far more promising.[27]

Work in Baden-Württemberg has not progressed as quickly or efficiently
as in Alsace-Lorraine. Three tributaries are potentially available for pisciculture: the Murg, Rench, and Kinzig (not to be confused with the identically named stream in Hesse). The Murg, once a favorite destination for
migratory fish, now has twenty-seven weirs separating its mouth from its

headwaters, where most of its remaining spawning and nursery grounds are located. The Rench has similar problems: twenty-six of the river's many channel barriers are impassable for adult fish. The Kinzig holds out the most immediate promise: its principal migration barriers—at Offenburg, Steinach, and Willstätt—have recently been retrofitted or eliminated, opening up some of its upstream gravel beds to migrants. Assuming the Baden-Württemberg government steps up its restoration efforts, all three streams may one day support self-sustaining populations.[28]

Salmon restoration in Switzerland poses the greatest number of constraints. As of now, hydrodams on the Upper Rhine render it all but impossible for any salmon to reach the base of the Alps at Basel. Yet even if they were one day to reach Basel, they would still find the hydrodams on the High Rhine (from Birsfelden to Schaffhausen). The Swiss government has so far restricted itself to small repopulation experiments on three tributaries in the vicinity of Basel—the Birs, Ergolz, and Wiese—in the expectation that the Upper Rhine's route will eventually be restored. As elsewhere, the preparatory work has consisted chiefly of identifying and restoring the best gravel beds, and removing in-stream hurdles. No effort has yet been made to retrofit the High Rhine's dams.[29]

Initially Salmon 2000 targeted only those Rhine tributaries and feeder streams that possessed well-oxygenated waters and a relative abundance of spawning and nursing sites—chiefly the Sieg, Saynbach, Lahn, Rench, and Ill. As researchers identified gravel beds elsewhere in the basin, they increased the number of targeted rivers to include two more lower tributaries (the Ruhr and Wupper), two middle tributaries (the Mosel and Main), the Upper Rhine's remnant bed, and three Alpine streams (Birs, Ergolz, Wiese). The prospects for success elsewhere in the Rhine basin, however, are poor: chemical contaminants, thermal pollution, overengineering, gravel extraction, and siltation all conspire against a cost-effective repopulation plan.

The Rhine Protection Commission estimates that there are approximately 150 hectares (370 acres) of suitable spawning grounds and 630 hectares (1556 acres) of smolting grounds left in the entire Rhine basin. A single female scatters, on average, 10,000 eggs over a space of one hundred square meters, with an egg survival rate of about one percent. Spawning space of 150 hectares will thus support 15,000 females depositing 15 million eggs and producing 1.5 million smolts. Nursery space of 630 hectares is sufficient to ensure that at least 600,000 of these smolts will survive long enough to make the journey downstream. Based on an average return rate of one to two percent, the Rhine could thus potentially support an annual run of 6000 to 12,000 thousand adult individuals—barely enough to ensure a self-sustaining population.[30]

So far, Salmon 2000 has failed to attain anything remotely close to those levels. In fact, the number of mature returning salmon has been extraordinarily small compared with the millions of eggs and parr introduced into the basin over the past decade. As of 1999, only 180 returning salmon had been spotted: 114 on the Sieg, 8 on the Wupper and Dhünn, 13 on the Saynbach, 1 on the Lahn, and 44 on the Ill and Bruche. Typically only five to ten percent of returning salmon are detected as they swim upstream, which suggests that somewhere between 1000 and 2000 adult salmon returned to their spawning grounds between 1989 and 1999—on average, 100 or 200 per year. Even more revealing: evidence of natural reproduction has been found so far only on the Sieg (since 1994) and Ill (since 1997).[31]

Salmon 2000 has nonetheless provided the skeletal structure upon which the flesh-and-blood of channel restoration can be laid in a remarkably short time. The return of salmon to the Sieg clearly demonstrates that Delta and Lower Rhine are now open to migration and spawning. The occasional appearance of salmon in the Ill suggests that migration routes are now slowly being extended to the Middle and Upper Rhine stretches as well. Salmon, moreover, are not the only gauge of success. Sea trout and lampern populations have begun to stabilize, and sea lamprey are mounting a slow comeback. Fintis shad and North Sea houting are seen sporadically on the Rhine and Meuse mouths, as are (on extremely rare occasions) allis shad. Between 1994 and 1997, for instance, Dutch fishermen snagged 113 fintis shad in the IJsselmeer, as well as 11 fintis shad and 2 allis shad in the Nederrijn-Lek and Waal. These numbers are not large compared with catches of the past, but they suggest that migratory populations are beginning to rebound after nearly a century of continuous decline.[32]

While Salmon 2000 has captured almost all the media attention, other projects have had a positive impact on Rhine ecology over the past two decades. One of the most innovative is the Stork Plan, jointly sponsored by the Dutch government and the World Wildlife Fund. As its name suggests, it has focused mostly on the restoration of habitat for the now-rare black stork, once plentiful in the delta's riverine forests and islands. The Stork Plan took an innovative approach to riverscape management by dividing the delta region into four zones: the main channel, where the needs of navigation and flood-control would continue to reign supreme; the river foreland (the corridor between the channel and dike), where nature protection was to become the paramount concern; the dike and levee corridors, which would continue to be used for a variety of purposes ranging from travel routes to horticultural use; and the basin areas (land behind the dikes and levees), where all dairy farming and agriculture would henceforth be located. In practice, the Stork Plan focused almost exclusively on the first two zones—the river channel and foreland—the areas most affected by previous engineering. The most extensive work has taken place on the Waal, the delta's main shipping lane between Pannerden and Rotterdam, with some modifications also occurring on the IJssel, Nederrijn-Lek, and Meuse. Most of the modifications were made at a few key locations on the delta's alluvial corridor—Millinger Waard, Sint Andries, Blauwe Kamer, and Duurse Waarden—in the hope that they would serve as the main centers of species renewal.[33]

The Stork Plan posed considerable engineering challenges to the Dutch water authorities. Nearly all previous river work had entailed the removal of alluvial forests and delta islands in order to increase water velocity and thus reduce the risk of backflow and overflow. Any restoration of in-channel forestland would inevitably have the opposite impact: it would impede the flow of water and thus augment the flood danger. Therefore, when engineers restored the alluvial forests at Sint Andries on the middle Waal, they added a lateral channel to the trunk river, in effect creating a new braid as an additional drainage route. Similarly, at Millinger Waard, a polder in the Gelderse Poort area downstream from the Pannerden canal, they deepened and widened the Waal in order to compensate for the slower flow anticipated by forest regeneration. Elsewhere on the Waal, portions of the river's clay strata were removed, sometimes in the interest of flood defense and sometimes to create a diverse habitat of pools and shallows that would encourage aquatic biodiversity.[34]

The Stork Plan established a clear demarcation between riverine and agricultural space on the Waal, with the main dikes and levees serving as the dividing line. That has obligated the Netherlands government to withdraw land on the river side of the dikes from agricultural production and to compensate farmers with arable soil on the land side. The plan also called for restoration of the Waal's original floodplain wherever possible. That has necessitated removing many summer dikes and other riverbank constructions that had previously constrained the river during highwater conditions.[35]

The French and German governments have also been experimenting with alluvial restoration on the Upper Rhine over the past fifteen years. In 1985, for instance, Baden-Württemberg removed dikes around the Altenheim polder just south of Strasbourg, reopening five square kilometers of former floodplain to periodic flooding. Between 1987 and 1992 alone, fifteen inundations occurred (on average three per year), enough to trigger a swift restoration of the region's morphological diversity, including riverbank, pool-riffle, island, and steep-face biotopes. As these niches were restored, the region's original biotic communities began returning, including kingfisher (a nesting bird), running and flying beetles, freshwater snails and bugs, and red alga. A rare water bug, *Aphelocheirus aestivalis*, found a new home in the polder's gravel riffs as well. Similar restoration work was undertaken in 1992 in Alsace at the Offendorf Nature Reserve (60 hectares), just north of Strasbourg.[36]

So far none of these habitat restoration projects have been large enough in scope to bring about a significant restoration of riverine plant and animal life, let alone a full rebound of salmon, stork, or water bugs. Nonetheless, they represent the first important steps in reestablishing the river's lost floodplain and alluvial corridor. Forestlands, gravel beds, islands, backwaters, and braids are all critically important riparian zones: their restoration is a prerequisite for greater biodiversity on the Rhine. Every hectare of restored gravel provides additional space not just for migratory fish, but for other animal communities, especially birds. The elimination of the river's superfluous dams and weirs, moreover, has allowed for a partial restoration of the Rhine's original hydrological and morphological features. This, in turn, has opened up new possibilities for bank stabilization, habitat variation, and species diversification in the basin.

Toward a New Flood Control Regime

In 1998, the riparian states signed a new agreement, the Convention for the Protection of the Rhine, which superseded the 1963 Bern Convention. The old convention had given the Rhine Protection Commission the mandate to reduce the river's thermal, chloride, and chemical loads. The new convention added two new tasks to the commission's agenda: ecological protection and flood control. For the most part, the new convention merely spelled out officially what had already become part of the commission's informal mandate. Flood control, however, was largely uncharted terrain.

The decision to add flood defense to the Rhine Protection Commisson's mandate came from the Rhine Ministers and the European Union. Much of the immediate impetus came from a succession of four "hundred-year" floods in the 1980s and early 1990s. Research into the causes of these floods confirmed what many had long suspected: that the various river engineering and land reclamation projects of the past two centuries had aggravated the flood danger on the Middle, Lower, and Delta Rhine. Salmon 2000, the Stork Plan, and similar projects also provided impetus: their success can be assured only if portions of the river's alluvial land and flood regime are restored. By linking water quality, flood control, and ecological protection, the 1998 Rhine Convention for the first time created the framework for a more holistic approach to river management.[37]

The "Rhine Action Plan on Flood Defence," endorsed by the Rhine Ministers in January 1998, empowered the Rhine Protection Commission to "compensate for the ecological deficits of the past" by removing "human interferences with the river regime, as far as possible." To accomplish this goal, the convention laid down several new principles of river management. First, all future channel engineering projects were to take cognizance of the impact on the entire river basin ("water is part of the natural ecological cycle of all surfaces and of land use and must be taken into account by all fields of policy"). Second, land reclamation was to come to an end, and in some cases be reversed: wherever possible agricultural fields were to be returned to floodplain. In areas where the floodplain had been permanently lost to cities and industries, polders (artificial storage basins) were to be constructed in the interest of flood defense. Third, all superfluous dams and overly high banks were to be removed from the basin ("let the river expand"). The Rhine and its tributaries were to be redesigned to allow the channels to overspill more readily during high-water periods, much as they would under natural conditions.[38]

Using 1995 as a base year, the Flood Defence Plan spelled out a twenty-five-year timetable for implementation. During Phase I (1995-2000) the Rhine Protection Commission was entrusted with compiling a comprehensive overview of flood-prone regions in the Rhine catchment basin. This task was largely accomplished with the publication of the *Rhine Atlas* in 1998, which identifies polder areas and maps sites where a return to natural conditions is economically feasible and ecologically necessary.[39] Phase II (2000-2005), now under way, is focused primarily on the establishment of water storage sites. In some regions, water storage can be accomplished simply by reopening selected portions of the former floodplain; in most regions, however, it will require artificial impoundment spaces (polders). The interim goal is to create enough spongy areas on the upper stretches of the Rhine to hold back 30 centimeters (12 inches) of water under peak conditions. In the long-term—Phase III (2005–2020)—the river's storage capacity is to be augmented to 70 centimeters (28 inches). A 70-centimeter drop in the river's high-water depth is sufficient to provide downstream protection under most circumstances.[40]

Because so much of the floodplain is now occupied by humans and human-based activities, only a partial return to natural conditions is feasible. Current plans foresee a restoration of approximately 160 square kilometers of the Rhine's former floodplain, along with a refashioning of approximately 1100 kilometers of its embankments. Within the entire Rhine basin, about 1000 square kilometers of former floodplain will be restored, and 11,000 kilometers of feeder streams. New agricultural policies will affect around 3900 square kilometers of the basin. Reforestation and nature development will affect another 3500 square kilometers. A series of polders—with storage capacities totaling 364 million cubic meters on the Rhine itself and an additional 73 million cubic meters on its tributaries—will function as the main mechanism for flood protection.[41]

Recognizing the limits of natural floodplain restoration, the Rhine Protection Commission initiated the Habitat Patch Connectivity Project at the same time that it announced its new flood-control measures. This project focuses on reestablishing the alluvial corridor that once stretched uninterrupted from Lake Constance to Hoek van Holland. The initial goal is to establish new natural and near-natural regions large enough in size to provide adequate living space for individual species, especially endangered ones. The next goal is to create as much habitat variation as possible on the river, including riverbank space for soft (willows) and hard (oak and elm) woods, and river-edge space for reeds and marshes. The ultimate goal, and the one that gave the project its name, is to reconnect these patches in order to provide migration and colonization paths for animals and plants. "Envisaged is a riverscape in which the most ecologically valuable and near-natural core areas are linked together so that every organism can reach every biotope," declared the Rhine Protection Commission, "for this will ensure the largest populations and the greatest variation. The Rhine's bed, bank, and floodplain are once again to become aquatic and terrestrial habitats that function as viable living space for animals and plants."[42]

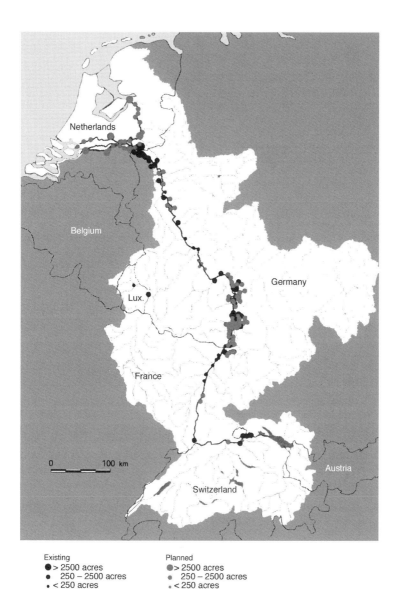

Existing
● > 2500 acres
● 250 – 2500 acres
· < 250 acres

Planned
● > 2500 acres
● 250 – 2500 acres
· < 250 acres

Figure 7.6 Map of the master plan for Rhine ecosystem restoration.
Floodplain renewal and restoration of natural conditions are occurring
on many stretches of the river. One hectare equals 2.47 acres. (Source:
Garritsen, Vonk, and de Vries, eds., *Visions for the Rhine*, 25)

Unfortunately, a full reestablishment of the river's once-thriving alluvial corridor is unlikely. All that will be possible in the foreseeable future is a partial reconstruction of the corridor. For instance, on the High Rhine (kilometer 0 at Lake Constance to kilometer 168 at Basel) there are today just three ecologically intact "core" stretches: from Stein am Rhein to Langwiesen/Paradies (kilometers 25 to 41.5), Rheinau to Tössreidern (kilometers 59.5 to 73.5), and Wallbach to Rheinfelden (kilometers 136.5 to 147.5). Collectively they account for only around one-fourth (41.5 kilometers) of the High Rhine's riparian corridor. Efforts are under way to restore three additional stretches: from Rheinfall to Rheinau (kilometer 50.2 to 59.5), Tössriedern to the Kaiserstuhl bridge (kilometer 73.5 to 83), and Reckingen to the Aare mouth (kilometer 90 to 102). Once completed, they will add another 30.8 kilometers to the tally, bringing the percentage of nature corridor to about 43 percent. But the remaining 57 percent of the corridor cannot be restored or reconnected to any significant degree because it has been fundamentally transformed by hydrodams and urban growth. "Reconnectivity" on the High Rhine will thus actually consist of a series of connected areas punctuated by sharp breaks.[43]

The prospects for reestablishing a river corridor on the Lower Rhine (kilometer 655 at Bonn to kilometer 867 at the Pannerden canal) are even worse, for there are today just a few small near-natural patches still extant. Between Wesel and Cleves (kilometers 814 to 857) in North Rhine–Westphalia, for instance, there are only around 20,000 hectares of unused space remaining, of which 11,000 are currently under protection. Efforts are under way to place about half of the remaining free space under protection and to add an additional 5000 hectares of ecologically valuable space nearby. In the stretch between the Dutch-German border and Pannerden (kilometers 862 to 867), there are nearly 8000 hectares of natural space, of which 3000 hectares are currently under protection. Over the next twenty-five years, the Dutch government intends to place all of this natural space under protection and add 8000 hectares of new land. Collectively, however, these projects will restore only a tiny fraction of the Lower Rhine's former floodplain—not even remotely enough to constitute a continuous river corridor.[44]

"For decades it was considered progressive to push for more and more land reclamation, but now it is also considered progressive to champion the right of wildness," Lauterborn wrote in 1903. "It has become imperative to save everything one possibly can that still retains the character of natural habitat."[45] Unfortunately, Lauterborn was well ahead of his time: it would take another half century before the Rhine riparian states would come to see value in preserving the Rhine of yesteryear, and even longer before they actively began to do something about it. The first big step came in 1963 when the Bern Convention breathed life into the new Rhine Protection Commission; the second came in 1986, when the Sandoz accident revealed just how precarious life had become for the Rhine's flora and fauna.

In 1869, the year of Lauterborn's birth, the Tulla Project was not yet complete, the chemical industry was in its infancy, and the salmon fisheries were in their heyday. By 1896, the year he completed his doctoral dissertation, the first of the High Rhine's hydrodams was under construction. While he served as head of the Palatinate's water research team from 1908 to 1945, the Upper Rhine's floodplain gradually disappeared as industries moved in and cities grew. As late as the 1920s, serious water pollution was confined mostly to the coal region of the Lower Rhine; the river's alluvial corridor was largely intact; and only a few native species were endangered. But by 1952, when Lauterborn died, the Rhine was well on its way to becoming a eutrophied river so denuded of vegetation and habitat that it had lost most of its former biodiversity. Not even a great biologist of Lauterborn's stature stood a chance against the Rhine Commission and its army of engineers. It was not until the river was on the brink of death—and therefore unable to fulfill its duties to humans—that a new attitude about the river began to emerge.

Concerted effort by the riparian states since 1963 has reversed the downward spiral, and the Rhine is now on its way to ecological recovery in three critical areas: water quality, biodiversity, and floodplain restoration. The greatest achievements so far have come in water quality. Though there are still some problem areas—notably the Emscher and Erft—water in the Rhine basin is considerably purer than it was during the peak pollution years of the mid-1970s. In the Alpine stretches, Class I/II is the rule, with the water only slightly compromised in purity by the presence of industries and cities. Even Lake Constance, which in 1979 had a phosphorus content three times higher than tolerable levels, is considerably cleaner than it was before.[46] The navigable Rhine is also much cleaner than it was thirty years ago, with Class II the norm in the upper stretches and Class II/III in the lower stretches. The chloride level has dropped significantly since 1992, much to the benefit of sensitive freshwater species. Two of the most problematic chemical classes, heavy metals and chlorinated hydrocarbons, have been brought under control. Even phosphorus and nitrogen levels have begun to decline, though nitrogen levels in particular are still a cause of concern. The quality of Rhine water will no doubt continue to improve as new mitigation technologies are developed and put in place, and the day may soon arrive when the entire Rhine will belong to Class I/II—about the best that an industrial river can hope to achieve.

Significant progress has also been made in reestablishing the Rhine's biodiversity. While the river no longer supports the full flora and fauna of yesteryear, there has nonetheless been a remarkable comeback both in the number and diversity of plant and animal species. Salmon, sea trout, sea lamprey, and lampern have all been mounting a modest comeback since the 1980s as the river's longitudinal and lateral pathways have slowly been reopened. Stationary fish species, too, have benefited from habitat restoration and better water quality, including the barbel, nase, and flounder—all once endangered or absent from their niches in the river. Roach and bream remain by far the most common fish, but the balance is slowly tipping in favor of other species. Diversity among invertebrates has also increased as oxygen levels have risen and chemical and chloride loads subsided. Some stretches of channel and bed now show a diversity of life not seen since 1900. Finally, some former floodplain has been restored, especially along those stretches (the Upper, Lower, and Delta Rhine) where the naturally soft and ever-changing banks once gave the river its expansive breadth. The extra riparian space has allowed for a modest rise in bird, amphibian, and mammal populations.

The full impact of the restoration work, of course, will not be visible for a long time to come. Floodplain forests reach their full maturity only after 250 years. Pioneer plants must first take root, then later come poplars and other light woods, then willows, then alders and ash, and finally elms and oaks.[47] Toxic buildup of heavy metals and micropollutant sediment deposits will also continue to affect riparian life long after the factories that produced them are closed. In the heavily polluted Biesbosch region of the Netherlands, for instance, tufted ducks still have trouble breeding, and night heron have yet to mount a comeback.[48] Nonetheless, the acorns of restoration have been planted.

These transformations are all the more remarkable because they represent a departure from the river regime that prevailed on the Rhine for over a century and a half. Until 1976, when the Rhine Ministers passed the first pollution conventions, the story of Rhine water quality was a story of decline. Since then, water purity has improved with each passing year. Similarly, until 1978 (when the last dam on the Upper Rhine was completed) river rectification invariably meant the loss of floodplain space. Since then, small portions of the floodplain have been restored at a few key locations, and more space will no doubt open up in the coming decades. It may even be possible to restore the river to something like the river that Lauterborn knew in his youth rather than the one he knew in his old age—a river with living space inside and outside its channel, with a diverse floodplain and corridor, and with animals and plants thriving amid the coalfields, chemical factories, hydrodams, and cities.

Who of us would not prefer to dwell
on the ingenious disorder of a natural river
landscape than on the trivial regularity
of a straightened water course?

— Friedrich Schiller (1793)

CHAPTER 8

Conclusion

*W*ater has given rise to many legends over the centuries, but none more famous than the "Sorcerer's Apprentice," the quintessential cautionary tale about unintended consequences. The story line dates back to Lucian of Samosata (circa A.D. 120–190), but its many nineteenth-century variants come from Goethe's poem *Der Zauberlehrling* (1797) and the fairy tales of the Brothers Grimm. In one popular version, a sorcerer, in possession of a magic broom, lived in a castle high above the Rhine. His apprentice, entrusted with the task of keeping the sorcerer's cistern full, hauled buckets of water each day up a steep and winding path from the river to the castle. One day, while the sorcerer was absent, the apprentice found a way to ease his burden: he invoked one of his master's spells and ordered the broom to haul the buckets for him. Relaxation turned to frenzy, however, when the apprentice belatedly realized that he only half understood the art of sorcery: he could command the broom to fill the cistern, but not to stop when the cistern was full. Soon the castle was bursting with river water. In desperation, he grabbed an ax and smashed the broom into small pieces—only to watch in horror as each of the splinters sprouted new arms and legs and began hauling water at an ever-faster pace.

The legend has endured because it rings so true: humans are easily seduced by solutions that promise a quick fix but end up delivering results laden with unforeseen perils. French revolutionaries set out to establish an international regime that would promote free commerce and trade on a river they instinctively knew was rich in economic potential. They had no intention of undermining the river's viability as a biological habitat, even though the one-dimensionality of their course of action led in that direction. Similarly, the Vienna diplomats placed Rhineland and Westphalia in Prussian hands primarily to keep France's eastward march on the Rhine in check. They did not intend to fuel German nationalism and militarism, or spark another round of Franco-German antagonism and another century of Rhine warfare. Tulla devoted most of his professional life to devising ways to protect the Upper Rhine's villages and towns from the ravages of Alpine flooding. He did not comprehend that his project would cause long-term erosion problems and exacerbate the downstream flood danger. Successive generations of engineers, industrialists, and urban planners set out to improve the river's navigability and thus make the riparian states more prosperous. They did not intend to kill the goose that laid the golden egg—to destroy its floodplain, pollute its water, and extirpate its oak and salmon. Collectively, they transformed the Rhine into one of the world's premier commercial streams, but at the same time turned it into a degraded biotope. Like powerful sorcerers, they put the Rhine to work for them. But like lowly apprentices, they had only half mastered the art of hydraulic engineering.

Today's engineers would not construct a navigational waterway the same way their predecessors did two hundred years ago. Ecologically based principles have come to assert themselves where techological principles once reigned supreme. Under pressure from biologists and ecologists, they appreciate the importance of preserving a river's living spaces as they manipulate its channel. The old rules of river engineering—based on the primacy of a single straight, wide, swift channel—have largely been discredited and abandoned. Curves, islands, backwaters, deep pools, river-edge habitat, longitudinal pathways, and floodplains are now preserved as much as possible.[1] Past engineers thought mechanically and designed geometrically; their "ideal river" had the straight and uniform profile of a canal, their ideal channel a trapezoidal or U-shape. Today's engineers take a different approach: they even build canals with biological principles in mind. The new Rhine-Main-Danube canal (1992), for all its monstrous artificiality, meanders through the Altmühl valley as if it were a natural river. Instead of cutting a straight line through the hills, engineers followed the contours of the valley landscape (after being forced to do so by the environmental lobby). They created river-edge habitats for aquatic and semiaquatic species, added artificial feeder streams to replicate the dynamics of a natural river, and paid attention to longitudinal pathways. As a result, the Rhine-Main-Danube canal looks more natural in many stretches than the three rivers it connects—and supports a surprising amount of plant and animal life for a wholly artificial and newly constructed waterway.[2]

Starting from scratch, however, is infinitely easier than undoing what has already been done. In the Rhine of legend, the sorcerer returned to the castle in the nick of time and saved the apprentice from his self-destructive path. Brooms, buckets, and water magically disappeared at the sorcerer's command, and castle life resumed as if nothing had gone awry. The Rhine of reality, however, cannot be restored with the wave of a magic wand. Renaissance and Enlightenment engineering principles are deeply embedded in today's Rhine in the form of concrete banks, hydrodams, locks, and similar in-stream structures. Removing them would endanger the network of cities, towns, villages, roads, and factories that have ensconced themselves in the riverscape. In the Rhine of legend, the apprentice was so chastened by his misadventure that he willingly resumed hauling river water by hand. In the Rhine of reality, the riparian states want a cleaner and healthier river—but one that continues to fulfill its assigned role in Europe's industrial-urban life. Any river that is used simultaneously for transportation, irrigation, power generation, industrial production, and urban sanitation has to be "harnessed." It cannot be allowed to flow as its own needs dictate. All restoration projects will therefore have to take place within the interstices of these interlocking human needs.

The Rhine Protection Commission and its affiliates have come up against these limits repeatedly over the past thirty years. The Salmon 2000 project will undoubtedly reestablish a self-sustaining salmon population in the Rhine, but coaxing a few salmon back to its channel is not the same as turning the Rhine into a true "salmon river" once again. The old river possessed enough fish habitat to support annual salmon runs of half a million or more. Today it is highly unlikely that the Rhine can sustain more than a tiny fraction of that number—perhaps twenty thousand salmon once the restoration efforts are complete. Shad and sturgeon may also mount a small recovery, but neither is likely to attain numbers that would support fishing families as in the past. Thermal pollution also limits restoration. Though every power plant today is obligated to restrict its heat emissions, these controls guarantee only that the Rhine will not continue to become warmer. There has been no serious discussion of allowing the river to return to its natural rhythm of freeze and thaw (a condition for which its original flora and fauna are well adapted) because ice blockages and ice floes would endanger year-round shipping. Similar problems beset alluvial restoration: humans have emplaced so many permanent fixtures on the Rhine's banks that there seems little likelihood that a continuous river corridor can be reestablished. There are no plans to restore the three thousand or more islands that once dotted the channel, because doing so would augment the flood danger and compromise boat safety. There are no plans to rebraid or remeander the channel to any significant degree, for that would make travel distances longer and more arduous. And there are no plans to fully reopen the river's longitudinal pathways, because removing the hydrodams would endanger Europe's energy supplies. Indeed, restoring the river to anything remotely like it was in 1815 would entail an intervention far beyond anything currently under consideration.

These constraints suggest that the Rhine will continue to look more like the Rhine of today than the Rhine of yesteryear, at least for the foreseeable future. But the long-term prospects for rehabilitation may be more favorable. Rivers, as Heraclitus once famously declared, are the incarnations of constant change ("You cannot step into the same river twice"). They are protean sculptors constantly mutating to fit their surroundings while inexorably wearing down and transforming everything in their paths. The Rhine, like all rivers, is in perpetual motion, simultaneously adapting to and challenging its constraints, relentlessly determined to escape the uniformity that humans have imposed upon it. Left to its own devices, it will eventually burst every dam, wash away every embankment, restore every bend, and recoup every hectare of floodplain, as if it possessed a memory of its former contours. The river's own efforts to free itself will force future generations to decide which dams are worth reinforcing and which are not; which embankments are worth keeping and which are not; which tributaries need to be harnessed and which do not. The European Union's agricultural policies, which are shifting some of the crop burden away from the Rhine basin, may also benefit the river's ecology over time. As more land goes out of production, the river will have more space to expand and be burdened with fewer agricultural contaminants. It took the better part of two centuries to construct the modern mechanical Rhine. It may take another two centuries to restore the biological Rhine.

"The Rhine is a providential river, but is also seems to be a symbolic one," Victor Hugo noted in 1838. "In its windings, in its course, amidst all that it traverses, it is the image of the civilization it has served and will continue to serve."[3] In his day the Rhine still had the upper hand, dictating to humans when and where they could safely construct their towns and villages, place their agricultural fields, and sail their boats. But even then, humans were increasingly taking control of the Rhine's hydraulic rhythms, dictating to the river where it would flow, what it would carry, and how its waters would be used. The transformation of the Rhine from a free-flowing to a harnessed river was slow and gradual—far too slow for most of those living on the river's banks to fully comprehend. But by the mid-1970s the river was so thoroughly manipulated, and its self-cleansing ability so compromised, that its survival as a biological habitat was seriously in doubt. In some cases, humans themselves sapped the river's vital lifelines by usurping its floodplain and banks. In other cases, humans acted as inadvertent catalysts of decline, as when they added phosphorus and nitrate nutrients to the river, creating a chain reaction that began with excess algal growth and ended with water eutrophication.

In Hugo's day, the Rhine was a magnet for Romantics like himself—the personification of all that was free, unique, quirky, unpredictable, willful, dark, brooding, historical, mystical, secretive, and moody. Once the engineers had remade it in their own "enlightened" image, the river gradually lost its allure and came to symbolize all that was wrong with industrial Europe. Recent restoration efforts have managed to reverse the downward spiral. The river is far cleaner, more spacious, and more biologically robust than it was thirty years ago. It may never inspire another generation of poets and painters, for the Romantic Rhine, like the Roman Rhine before it, now belongs to the past. But as the river comes back to life, it has at least begun—at long last—to shed its image as Europe's sewer.

Notes

Chapter 1. Introduction

1. Johann Gottfried Tulla, "Bericht an das Großherzogliche Ministerium der auswärtigen Angelegenheiten, vom 1.3. 1812." Cited by Max Honsell, *Die Korrektion des Oberrheins von der Schweizer Grenze unterhalb Basel bis zur Großherzogthum Hessischen Grenze unterhalb Mannheim* (1885), 5.

2. Cited by Edwin J. Clapp, *The Navigable Rhine* (1911), 6. The stanza in German reads "Mehr Zölle sind am Rhein als Meilen, Und Pfaff' und Ritter sperrt den Strom." Clapp translates it more poetically as "The Rhine can count more tolls than miles, and knight and priestling grind us down."

3. Richard White, *The Organic Machine: The Remaking of the Columbia River* (1995).

4. Julius Caesar, *The Gallic War* (1963), book IV:10.

5. Ann Radcliffe, *A Journey, Made in the Summer of 1794 Through Holland and the Western Frontier of Germany with a Return Down the Rhine* (1795), 55.

6. Henry David Thoreau, "Walking," in *The Works of Henry D. Thoreau: Excursions* (1913), 175–76.

7. Joost van den Vondel, "De Rynstroom," *De Werken van Vondel*, vol. 3 (1929), 289.

8. Cited by Cecelia Hopkins Porter, *The Rhine as Musical Metaphor: Cultural Identity in German Romantic Music* (1996), 34.

9. Friedrich Hölderlin, *Hymns and Fragments* (1984), 71.

10. Lord Byron, *Childe Harold's Pilgrimage and Other Romantic Poems* (1936 [1819]), 104.

11. Heinrich Heine, "Ein Märchen aus alten Zeiten" (1827), reprinted in Helmut J. Schneider, ed., *Der Rhein: Seine poetische Geschichte in Texten und Bildern* (1983), 199–200.

12. Victor Hugo, *Le Rhin* (1980 [1845]), 142.

13. International Commission for the Hydrology of the Rhine Basin (CHR), *Der Rhein unter der Einwirkung des Menschen—Ausbau, Schiffahrt, Wasserwirtschaft* (1993), 23.

14. Cited by Hans Boldt, "Deutschlands hochschlagende Pulsader," in Hans Boldt, ed., *Der Rhein: Mythos und Realität eines europäischen Stromes* (1988), 30.

15. Albert Demangeon and Lucien Febvre, *Le Rhin: Problèmes d'histoire et*

d'économie (1935).

16. Cited by Boldt, "Deutschlands hochschlagende Pulsader," in Boldt, ed., *Der Rhein*, 31.

17. Heinrich von Kleist, "Germanias Aufruf an Ihre Kinder," in Ilse-Marie Barth, Klaus Müller-Salget, Stefan Ormanns, and Hinrich C. Seeba, eds., *Erzählungen, Anekdoten, Gedichte, Schriften* (1990), 427–33. The poem was published in 1813, two years after his death.

18. The "poets' war" is described in detail by Porter, *The Rhine as Musical Metaphor*, 38–47. Not all Germans, of course, got caught up in the doggerel. In Heinrich Heine's lampoon, Father Rhine said of Becker's poem: "When I hear it, that dumb song/ It makes me want to tear off my white beard / And drown myself in me."

19. Horst-Johannes Tümmers, *Rheinromantik: Romantik und Reisen am Rhein* (1968), 15.

20. Torsten Mick and Michael Tretter, "Der Rhein und Europa," in Boldt, ed., *Der Rhein*, 43.

21. Werner Reh, "Ökologische Folgen der Industrialisierung des Rheins," ibid., 117–19.

22. Ragnar Kinzelbach, ed., *Die Tierwelt des Rheins einst und jetzt: Symposium zum Jubiläum der Rheinischen Naturforschenden Gesellschaft und des Naturhistorischen Museums Mainz am 9. November* 1984 (1985), 40.

23. Franz Xaver Michels, "Die Entstehungsgeschichte des Rheins," *Beiträge zur Rheinkunde*, no. 25 (1973): 3–24. See also Kinzelbach, ed., *Die Tierwelt des Rheins einst und jetzt*, 7–25; Günther Reichelt, *Laßt den Rhein leben!* (1986), 7–9; and Anton Lelek and Günter Buhse, *Fische des Rheins—früher und heute* (1992), 3.

24. Rheinstrombaudirektor Langen, "100 Jahre Rheinstrombauverwaltung Koblenz," *Schriftenreihe des Rhein-Museum e.v. Koblenz* (1950), 5.

25. Garrett Hardin, "The Tragedy of the Commons," *Science* 162 (1968): 1243–48.

26. Lewis Mumford, *Technics and Civilization* (1934), 156.

27. Hartmut and Antje Solmsdorf, "Schutzwürdigen Bereiche im Rheintal," *Beiträge zur Rheinkunde*, no. 27 (1975): 28–31.

28. *The Complete Poetical Works of Samuel Taylor Coleridge* (1957), vol. 1, 477. Later entitled "Cologne," it first appeared in 1834 as part of a poem entitled "Lightheartednesses in Rhyme," in *Friendship's Offering*.

Chapter 2. Europe's "World River"

1. Manfred Fenzl, *Der Rhein: Schaffhausen-Nordsee und zum IJsselmeer* (1994), 8–9.

2. CHR, *Der Rhein unter der Einwirkung des Menschen*, 13. If the Meuse basin is included, the Rhine's catchment area becomes 225,000 square kilometers.

3. Egbert Wever, "The Port of Rotterdam: 'Gateway to Europe'," in Heinz

Heineberg, Norbert de Lange, and Alois Mayr, eds., *The Rhine Valley—Urban, Harbour and Industrial Development and Environmental Problems: A Regional Guide Dedicated to the 28th International Geographical Congress, The Hague 1996* (1996), 13. Rotterdam achieved the number one ranking from the New York Port in 1962. It has maintained its status up to today, but it seems only a matter of time before Singapore will overtake it.

4. Fenzl, *Der Rhein*, 9.

5. CHR, *Der Rhein unter der Einwirkung des Menschen*, 26; and Elmar Sabelberg, "Die Städteballungen am Rhein," in Boldt, ed., *Der Rhein*, 139–46.

6. On the chemical industry, see Fred Aftalion, *A History of the International Chemical Industry* (1991), 375. He ranked Bayer, Hoechst, and BASF first, second, and third as of 1988, Ciba-Geigy ninth, Sandoz twenty-second, and Hoffmann–La Roche twenty-ninth.

7. "Der Rhein, ein europäisches Umweltproblem," *Aktuelle JRO-Landkarte* (1985).

8. CHR, *Der Rhein unter der Einwirkung des Menschen*, 23. The CHR includes Italy as a riparian power because a minuscule amount of the Rhine's Alpine watershed lies in northern Italy.

9. The first attempt to subdivide the river was made by C. Mordziol, "Zur Gliederung des Rheinstromes in einzelne Abschnitte," *Geographischer Anzeiger* 13 (1912): 231–32. Subsequent researchers have modified Mordziol's divisions, without however creating a standard accepted by all. For a useful discussion of the subdivisions, see G. Friedrich and D. Müller, "Rhine," in B. A. Whitton, ed., *Ecology of European Rivers* (1984), 265–315.

10. Goronwy Rees, *The Rhine* (1967), 32–38. Statistical data from International Commission for the Hydrology of the Rhine Basin (CHR), *Das Rheingebiet—Hydrologische Monographie* (1977).

11. Lelek and Buhse, *Fische des Rheins*, 4.

12. *La Navigation fluviale: Edition Spéciale de la Revue Schweizerland* (July 1918): 19–20.

13. CHR, *Der Rhein unter der Einwirkung des Menschen*, 16.

14. CHR, *Das Rheingebiet*, Teil A (Texte), 107–52.

15. CHR, *Der Rhein unter der Einwirkung des Menschen*, 18.

16. Ibid., 17.

17. Compiled from CHR, *Das Rheingebiet*, Teil A (Texte), 131; CHR, *Der Rhein unter der Einwirkung des Menschen*, 38; and Deutsche Forschungsgemeinschaft, *Hydrologischer Atlas der Bundesrepublik Deutschland* (1979), 147.

18. Ton Garritsen, Guido Vonk, and Kees de Vries, eds., *Visions for the Rhine* (2000), 31.

19. Traude Löbert, *Die Oberrheinkorrektion in Baden: Zur Umweltgeschichte des 19. Jahrhunderts* (1997), 14.

20. See International Commission for the Hydrology of the Rhine Basin (CHR), *Die Hochwasser an Rhein und Mosel im April und Mai 1983* (1989) and *Das Hoch-*

wasser 1988 im Rheingebiet (1990). See also Bundesanstalt für Gewässerkunde, *Das Hochwasser 1993/94 im Rheingebiet* (1994).

21. Article 5 of the 1814 Paris Peace Treaty, and Articles 108–116 and Annex XVI B of the 1815 Vienna Final Acts.

22. *Kölnische Zeitung* (13 June 1816).

23. See Dietrich Ebeling, *Der Holländerholzhandel in den Rheinlanden: zu den Handelsbeziehungen zwischen den Niederlanden und dem westlichen Deutschland im 17. und 18. Jahrhundert* (1992).

24. Cesare S. Maffioli, *Out of Galileo: The Science of Waters, 1628–1718* (1994), 268; and Norman Smith, *Man and Water: A History of Hydro-Technology* (1975), 43–45.

25. André E. Guillerme, *The Age of Water: The Urban Environment in the North of France, A.D. 300–1800* (1983), 196–209 (quote 207).

26. Johann Gottfried Tulla, *Denkschrift: Die Rektifikation des Rheines* (1822), cited by Heinz Musall, *Die Entwicklung der Kulturlandschaft der Rheinniederung zwischen Karlsruhe und Speyer vom Ende des 16. bis zum Ende des 19. Jahrhunderts* (1969), 197.

27. Cited by H. Wittmann in "Tulla, Honsell, Rehbock," *Bautechnik-Archiv*, no. 4 (1949): 12.

28. For these and other instances of militarized language, see Löbert, *Die Oberrheinkorrektion in Baden*, 98–99.

29. See the *procès verbaux* of the Congress of Vienna's Commission on the Free Navigation of Rivers, *Rheinurkunden: Sammlung zwischenstaatlicher Vereinbarungen, landesrechtlicher Ausführungsverordnungen und sonstiger wichtiger Urkunden über die Rheinschiffahrt seit 1803*, vol. 1, (1918), 55–162.

30. Pierre Ayçoberry and Marc Ferro, *Une histoire du Rhin* (1981), 370–71. The "international river" idea had been circulating in diplomatic circles since the Peace of Westphalia (1648), but had never been implemented.

31. Article 5 of the Paris Peace Treaty (1814), in *Rheinurkunden*, vol. 1, 36.

32. The phrase *jusqu'à son embouchure* comes from the Vienna Final Acts (1815), Article 109; see *Rheinurkunden*, vol. 1, 42–43. For the Dutch position, see H. P. H. Nusteling, *De Rijnvaart in het tijdperk van stoom en steenkool (1831–1914)* (1974), 5–9.

33. The texts of the Mainz Acts and Mannheim Acts are in *Rheinurkunden*, vol. 1, 213–83, and vol. 2, 80–106. See also J. H. Schawacht, *Schiffahrt und Güterverkehr zwischen den Häfen des deutschen Niederrheins (insbesondere Köln) und Rotterdam vom Ende des 18. bis zur Mitte des 19. Jahrhunderts (1794–1850/51)* (1973), 11–25. For a detailed analysis of the diplomacy from 1815 to 1866, see J. P. Chamberlain, "The Regime of the International Rivers: Danube and Rhine," (1923); and A. Seeliger, "Die internationale Rechtsordnung des Rheins," in Walter Schmitz, ed., *50 Jahre Rhein-Verkehrs-Politik* (1927), 115–51.

34. The Rhine Commission's history can be found in Ursula von Köppen, "Die Geschichte der Kommission und die Rechtsordnung der Rheinschiffahrt," in *150*

Jahre Zentralkommission für die Rheinschiffahrt (1966), 21–28; and Kurt Lenzner, "Die internationale Rechtsordnung des Rheines," in Wasser- und Schiffahrtsdirektion Duisburg, ed., *Der Rhein: Ausbau, Verkehr, Verwaltung* (1951), 389–99. The United States replaced Italy as a member in 1963.

35. Dethard Freiherr von dem Bussche-Haddenhausen, "Einiges über Geschichte und Tätigkeit der Zentralkommission für die Rheinschiffahrt," *Beiträge zur Rheinkunde*, no. 19 (1968): 13–29.

36. Clapp, *The Navigable Rhine*, 17–19.

37. James C. Scott, *Seeing Like a State: How Certain Schemes to Improve the Human Condition Have Failed* (1998).

38. The documents relating to 1816–1916 are in *Rheinurkunden*, vols. 1 and 2; the documents on the transport of hazardous goods are all in vol. 2. (for 1916–1966). See also Köppen, "Die Geschichte der Kommission," 21–28.

Chapter 3. Water Sorcery

1. Publius Cornelius Tacitus, *The Annals* (1955), book XIII:53.

2. Cited by Zvi Herman, *The River and the Grain* (1988), 37.

3. Boldt, "Deutschlands hochschlagende Pulsader," in Boldt, ed., *Der Rhein*, 27–34.

4. CHR, *Der Rhein unter der Einwirkung des Menschen*, 42–44.

5. Henk Meijer, "The Dutch River District," in Heineberg, Lange, and Mayr, eds., *The Rhine Valley*, 22; and A. C. de Gaay, *The Canalization of the Lower Rhine* (1970), 5.

6. Langen, "100 Jahre," 6; and CHR, *Der Rhein unter der Einwirkung des Menschen*, 40–42.

7. CHR, *Der Rhein unter der Einwirkung des Menschen*, 41. See also Christine Hoppe, *Die großen Flußverlagerungen des Niederrheins in den letzten zweitausend Jahren und ihre Auswirkungen auf Lage und Entwicklung der Siedlungen* (1970), which offers a detailed analysis of channel changes over two thousand years on the stretch near the contemporary German-Dutch border.

8. For a succinct biography, see Martin Eckoldt, "Johann Gottfried Tulla—Zu seinem 200. Geburtstag," *Beiträge zur Rheinkunde*, no. 22 (1970): 19–22. See also Heinrich Cassinone and Karl Spieß, *Johann Gottfried Tulla: Sein Leben und Wirken* (1929); K. Knäble, "Tätigkeit und Werk Tullas," *Badische Heimat: Mein Vaterland* 50, no. 4 (1970): 450–65; Egon Kunz, "Von der Tulla'schen Rheinkorrektion bis zum Oberrheinausbau," *Jahrbuch für Naturschutz und Landschaftsplege* 24 (1975): 59–78; and H. G. Zier, "Johann Gottfried Tulla: Ein Lebensbild," *Badische Heimat* 50 (1970): 379–449.

9. Ernst-Volker Bärthel, *Der Stadtwald Breisach: 700 Jahre Waldgeschichte in der Aue des Oberrheins* (1965), 12.

10. J. G. Tulla, *Über die Rektifikation des Rheins, von seinem Austritt aus der*

Schweiz bis zu seinem Eintritt in das Großherzogthum Hessen (1825), 9.

11. Honsell, *Die Korrektion des Oberrheins*, 31–32.

12. Rolf Gustav Haebler, *Badische Geschichte* (1951), 111.

13. Eckoldt, "Johann Gottfried Tulla," *Beiträge zur Rheinkunde*, no. 22 (1970): 19–22.

14. Wittmann, "Tulla," *Bautechnik-Archiv*, no. 4 (1949): 11–12; Herbert Schwarz-mann, "War die Tulla'sche Oberrhein-korrektion eine Fehlleistung im Hinblick auf ihre Auswirkungen?" *Die Wasserwirtschaft* 54 (1964): 279–82; Kunz, "Von der Tulla'schen Rheinkorrektion bis zum Oberrheinausbau," 59–67.

15. Löbert, *Die Oberrheinkorrektion in Baden*, 67–80; and Christoph Bernhardt, "Zeitgenössische Kontroversen über die Umweltfolgen der Oberrheinkorrektion im 19. Jahrhundert," *Zeitschrift für die Geschichte des Oberrheins* (1998) 296–99.

16. Karl Felkel, "Das Problem der Sohlenstabilisierung des Oberrheins und die Naturversuche mit Geschiebezugabe," *Beiträge zur Rheinkunde*, no. 33 (1981): 20–35; and Schwarzmann, "War die Tulla'sche Oberrhein-Korrektion eine Fehl-leistung?" 279–87.

17. The Tulla Monument is on the right bank of the Rhine at kilometer 361.

18. Gerhard Mantz, "Zur Erinnerung an Leben und Werk des Geheimen Regierungsrathes und Strombaudirektors Eduard Adolph Nobiling," *Beiträge zur Rheinkunde*, no. 34 (1982): 22–38 (quotation 26).

19. Ibid.

20. Langen, "100 Jahre," 9; and CHR, *Der Rhein unter der Einwirkung des Men-schen*, 130.

21. Karl Felkel, "Strombau-Geschichte der Binger-Loch-Strecke des Rheins," and Karl Pichl, "Die Verbesserung des Schiffahrtsweges in der Binger-Loch-Strecke," both in *Beiträge zur Rheinkunde*, no. 12 (1961): 26–44 and 45–59.

22. Mantz, "Zur Erinnerung," 35–36.

23. Kurt Tucholsky, "Denkmal am Deutschen Eck," in Schneider, ed., *Der Rhein*, 383.

24. G. P. van de Ven, ed., *Man-Made Lowlands: History of Water Management and Land Reclamation in the Netherlands* (1993), 227–28.

25. Ibid., 229–30.

26. Ibid., 232–33.

27. Ibid., 233–34. See also Ben Wiebenga, "Pieter Calands Plan," *Beiträge zur Rheinkunde*, no. 24 (1972): 19–24; Johan van Veen, *Dredge, Drain, Reclaim: The Art of a Nation* (1962), 80–105.

28. CHR, *Der Rhein unter der Einwirkung des Menschen*, 188–90.

29. The diplomacy is handled in detail by Fritz Koenig, *Die Verhandlungen über die internationale Rheinregulierung im st. gallisch-voralbergischen Rheintal von den Anfängen bis zum schweizerisch-österreichischen Staatsvertrag von 1892* (1971); and Thomas Kuster, "Die Geschichte der internationalen Rheinregulierung und ihre Auswirkung auf die Bevölkerung und Wirtschaft des Vorarlberger Rheintals" (1991).

30. Koenig, *Die Verhandlungen über die internationale Rheinregulierung*, 17–20. See also W. Versell and A. Schmid, *Bericht über Wildbachverbauungen im bündnerischen Rheingebiet zur Sicherung der Rheinregulierung oberhalb des Bodensees* (1928).

31. Ferdinand Waibel, "Die Werke der Internationalen Rheinregulierung," in *Der Alpenrhein und seine Regulierung: Internationale Rheinregulierung, 1892–1992* (1992), 206–35.

32. Georg Weber and Romano Zgraggen, "Schweizerische Kraftwerke im Rhein-Einzugsgebiet," in *Der Alpenrhein*, 355–59. See also Herbert Calvis, "Die wasser- und energiewirtschaftliche Bedeutung des Rheins von seinen Quellen bis zum Eintritt ins Rheinische Schiefergebirge" (1981), 22–99. Industrial development on this stretch will be handled in chapter 5.

33. Niklaus Schnitter, *Geschichte des Wasserbaus in der Schweiz* (1992), 117–20.

34. Oberregierungsbaurat Altmayer, "Vom Oberrhein zwischen Basel und Bodensee," *Der Schaffende Rhein: Neue Folge der Beiträge der Rheinfreunde*, no. 7 (1931): 4–22; H. R. Strack, "Der Hochrhein von Basel bis zum Bodensee," *Beiträge zur Rheinkunde*, no. 11 (1960): 27–37; Hugo Ott and Thomas Herzig, "Elektrizitätsversorgung von Baden, Württemberg und Hohenzollern 1913/14," *Historischer Atlas von Baden-Württemberg* (1972–88), vol. 11, 9; and Calvis, "Die wasser- und energiewirtschaftliche Bedeutung des Rheins," 220–302. Industrial development on this stretch will be handled in chapter 5.

35. H. Bertschinger, "Schiffbarmachung und Wasserkraftverwertung des Rheins von Straßburg bis Basel," in Max Fenner, ed., *Die Binnenschiffahrt und Wasserkraftnutzung der Schweiz* (1926), 97.

36. Wittmann, "Honsell, Rehbock," *Bautechnik-Archiv*, no. 4 (1949): 25–42.

37. Wasser- und Schiffahrtsdirektion Duisburg, *Der Rhein: Ausbau*, 115–25.

38. "Convention du 10 mai 1879 entre la Suisse et la Grand-Duché de Bade au sujet de la navigation sur le Rhin, de Neuhausen jusqu'en aval de Bâle," *Les Actes du Rhin et de la Moselle* (1966), 17.

39. *De la Suisse à la Mer: Edition Spéciale de la Revue Schweizerland* (September 1920).

40. J. Dieterlen, "Kembs: Premier Échelon du Grand Canal d'Alsace," *La Navigation du Rhin* 10 (November 1932): 405–69; and CHR, *Der Rhein unter der Einwirkung des Menschen*, 80–92.

41. Alwin Seifert, "Die Versteppung Deutschlands," *Deutsche Technik* 4 (September 1936): 423–27. For a good summary of Baden's genuine concerns, see Badischer Landwirtschaftlicher Hauptverband, *Steppe am Oberrhein? Der französische Rheinseitenkanal* (1954).

42. CHR, *Der Rhein unter der Einwirkung des Menschen*, 80–92.

43. Fritz André, *Bemerkungen über die Rectification des Oberrheins und Schilderung der furchtbaren Folgen, welche dieses Unternehmen für die Bewohner des Mittel- und Unterrheins nach sich ziehen wird* (1828), 17–19. For Freiherr van der Wijck's critique, see Bernhardt, "Zeitgenössische Kontroversen," 301–3.

44. For a complete analysis of the proceedings, see Cassinone and Spieß, *Johann Gottfried Tulla*, 47–67; Honsell, *Die Korrektion des Oberrheins*, 12–18; Löbert, *Die Oberrheinkorrektion in Baden*, 40–48; and Bernhardt, "Zeitgenössische Kontroversen," 299–306.

45. Honsell, *Die Korrektion des Oberrheins*, 12.

46. Ibid., 12–18.

47. Ibid., 15.

48. Bernhardt, "Zeitgenössische Kontroversen," 311.

49. Max Honsell, *Die Hochwasser-Katastrophen am Rhein im November und Dezember 1882* (1883), esp. 5–11 and 13–23.

50. J. H. Schawacht, *Schiffahrt und Güterverkehr*, 60.

51. Fenzl, *Der Rhein*, 9.

52. Clapp, *The Navigable Rhine*, 34.

53. The 1825 figures are from Christian Eckert, *Rheinschiffahrt im XIX. Jahrhundert* (1900), 154. The 1978 figures are from Commission Centrale pour la Navigation de Rhin, *Rapport Annuel de la Commission Centrale pour la Navigation du Rhin* (1985), 53.

54. Kaiserliches Statistisches Amt, *Statistisches Jahrbuch für das Deutsche Reich* (1880), 105.

55. Helmut Betz, *Die großen Motorschlepper und die Entwicklung der Schubschiffahrt auf dem Rhein* (1988), 5–9; and Heinz Weber, *Die Anfänge der Motorschiffahrt im Rheingebiet* (1978).

56. Friedbert Barg and Sandro Cambruzzi, *Schubeinheiten und Koppelverbände* (1991), 16–18 and 97–99; and Werner Böcking, "Die Entwicklung der Motorschiffahrt auf dem Rhein," in Ulrich Löber, ed., *2000 Jahre Rheinschiffahrt* (1991), 113–18; and Wirth, "The Rhine as a Shipping Route," in Heineberg, Lange, and Mayr, eds., *The Rhine Valley*, 62–63.

Chapter 4. The Carboniferous Rhine

1. Mumford, *Technics and Civilization*, 157.

2. "Nasmyth at Coalbrookdale," in Raymond Williams, ed., *The Pelican Book of English Prose* (1969), vol. 2, 154.

3. Mumford, *Technics and Civilization*, 165.

4. Jürgen Büschenfeld, *Flüsse und Kloaken: Umweltfragen im Zeitalter der Industrialisierung (1870–1918)* (1997), 203.

5. On the riparian cooperatives, see Allen V. Kneese and Blair T. Bower, "Die Wasserwirtschaft im Ruhrgebiet: Eine Fallstudie der Genossenschaften," in Horst Siebert, ed., *Umwelt und wirtschaftliche Entwicklung* (1979), 351–72. See also August Heinrichsbauer, *Industrielle Siedlung im Ruhrgebiet in Vergangenheit, Gegenwart und Zukunft* (1936), 35–56.

6. Heiner Jansen, "Bergbau und Gewerbe," in Helmut Hahn and Wolfgang

Zorn, eds., *Historische Wirtschaftskarte der Rheinlande um 1820* (1973), 43–52. See also Joachim Radkau, *Technik in Deutschland: Vom 18. Jahrhundert bis zur Gegenwart* (1989), 74–87. "Poetically agricultural" was T. C. Banfield's description of Essen in the 1840s, as cited by N. J. G. Pounds, *An Historical Geography of Europe* (1990), 418.

7. Statistics from "Die Entwicklung der Rhein-Ruhrhäfen und ihre Beziehung zur wirtschaftlichen Erschliessung des niederrheinisch-westfälischen Industriegebiets," NRW/HSA, Regierung Düsseldorf, Nr. 54009.

8. Friedrich-Carl Schultze-Rhonhof, *Die Verkehrsströme der Kohle im Raum der Bundesrepublik Deutschland zwischen 1913 und 1957: Eine wirtschaftsgeographische Untersuchung* (1964), 7; and International Energy Agency, *International Coal Trade: The Evolution of a Global Market* (1997), 101.

9. Statistics from "Die Entwicklung der Rhein-Ruhrhäfen und ihre Beziehung zur wirtschaftlichen Erschliessung des niederrheinisch-westfälischen Industriegebiets," NRW/HSA, Regierung Düsseldorf, Nr. 54009. See also Walter Steitz, "Eisenbahn und Steinkohlenbergbau im Ruhrgebiet," in Hans-Jürgen Teuteberg, ed., *Westfalens Wirtschaft am Beginn des "Maschinenzeitalters"* (1988), 317–35.

10. David S. Landes, *The Unbound Prometheus: Technological Change and Industrial Development in Western Europe from 1750 to the Present* (1969), 89–95.

11. Pounds, *An Historical Geography of Europe*, 408–9.

12. Rainer Fremdling, "Standorte und Entwicklung der Eisenindustrie," in Teuteberg, ed., *Westfalens Wirtschaft*, 297–316 (statistics, 308 and 312).

13. Pounds, *An Historical Geography of Europe*, 418–21.

14. Karl Overbeck, "Die Wanderung der Großeisenindustrie des Ruhrgebiets zum Rhein" (1923), 24–35. See also Fremdling, "Standorte und Entwicklung der Eisenindustrie"; Fremdling, "Eisen, Stahl und Kohle," in Hans Pohl, ed., *Gewerbe- und Industrielandschaften vom Spätmittelalter bis ins 20. Jahrhundert* (1986), 347–70.

15. Overbeck, "Die Wanderung der Großeisenindustrie," 56–59 and 94.

16. Statistics from Thomas Kluge and Engelbert Schramm, *Wassernöte: Zur Geschichte des Trinkwassers* (1988), 156.

17. Heinrichsbauer, *Industrielle Siedlung im Ruhrgebiet*, 19.

18. Ulrike Klein, "Die Gewässerverschmutzung durch den Steinkohlenbergbau im Emschergebiet," in Teuteberg, ed. *Westfalens Wirtschaft*, 341.

19. Thomas Rommelspacher, "Das natürliche Recht auf Wasserverschmutzung," in Franz-Josef Brüggemeier and Thomas Rommelspacher, eds., *Besiegte Natur: Geschichte der Umwelt im 19. und 20. Jahrhundert* (1987), 57–58.

20. Klein, "Die Gewässerverschmutzung," 342–44.

21. Rommelspacher, "Das natürliche Recht," 60–61.

22. Cited in Franz-Josef Brüggemeier, "The Ruhr Basin 1850–1980: A Case of Large-Scale Environmental Pollution," in Peter Brimblecombe and Christian Pfister, eds., *The Silent Countdown: Essays in European Environmental History* (1990), 212.

23. Cited by Kluge and Schramm, *Wassernöte*, 102.

24. Alexander Ramshorn, "Die Wasserwirtschaft im rheinisch-westfälischen Industriegebiet," *Glückauf* (Sonderdruck) 88 (1952): 9.

25. Thienemann's report is cited by Kluge and Schramm, *Wassernöte*, 174–76.

26. Ibid., 160.

27. Ibid., 159–60; and Peter Franke and Wolfgang Frey, *Talsperren in der Bundesrepublik Deutschland* (1987), 386.

28. Figures from Franke and Frey, *Talsperren*, 386–401.

29. Kluge and Schramm, *Wassernöte*, 168.

30. Ulrich Schenck, "Die Wasserwirtschaft im Niederschlagsgebiet der Ruhr: Eine volkswirtschaftliche Untersuchung" (1931), 28.

31. Heinrichsbauer, *Industrielle Siedlung im Ruhrgebiet*, 12–13, 19, and 58.

32. Untitled report in NRW/HSA, Regierung Düsseldorf, Nr. 55904.

33. D. A. Ramshorn, *Die Emschergenossenschaft* (1957), 5–35. See also Heinrichsbauer, *Industrielle Siedlung im Ruhrgebiet*, 10–13.

34. Klaus-Georg Wey, *Umweltpolitik in Deutschland: Kurze Geschichte des Umweltschutzes in Deutschland seit 1900* (1982), 77–86; and Kluge and Schramm, *Wassernöte*, 101–4.

35. Kluge and Schramm, *Wassernöte*, 180–81.

36. Cited by Rommelspacher, "Das natürliche Recht," 57. Similarly, Heinrichsbauer stated: "It was advantageous to drain as much of the Ruhr's huge supply of wastewater directly into the Rhine because the Rhine is particularly well suited to absorb wastewater and because the Rhine waterworks use a filter system to draw their water through the Rhine gravel banks and will therefore not be negatively impacted." Cited by Kluge and Schramm, *Wassernöte*, 179.

37. Ramshorn, *Die Emschergenossenschaft*, 51.

38. Kluge and Schramm, *Wassernöte*, 103.

39. Ramshorn, *Die Emschergenossenschaft*, 46–51.

40. "Vorläufige Denkschrift über die Phenolfrage des Oelbachgebiets im Zusammenhang mit den Ruhrwasserwerken" (30 December 1925), in NRW/HSA, Regierung Düsseldorf, Nr. 48741.

41. Cited by Rommelspacher, "Das natürliche Recht," 57.

42. August Heinrichsbauer, *Die Wasserwirtschaft im rheinisch-westfälischen Industriegebiet* (1936), 85–94.

43. Ibid., 95–106.

44. Wasser- und Schiffahrtsdirektion Duisburg, *Der Rhein: Ausbau*, 437–41.

45. Götz Voppel, "Rhenish Lignite Mining—Development and Regional Effects," in Heineberg, de Lange, and Mayr, eds., *The Rhine Valley*, 109–11. A century of intensive mining has hardly made a dent in the North Rhine Field: at current extraction rates (100 million metric tons per year) it will take another 400 years to deplete the remaining 39 billion metric tons of available lignite.

46. Schultze-Rhonhof, *Die Verkehrsströme der Kohle*, 19–22.

47. Voppel, "Rhenish Lignite Mining," 112–13.

48. Ibid., 110. See also Hans Pohl, *Vom Stadtwerk zum Elektrizitätsgroßunternehmen. Gründung, Aufbau und Ausbau der "Rheinisch-Westfälischen Elektrizitätswerk AG" (RWE) 1898–1918* (1992), especially 21–36.

49. Figures from Voppel, "Rhenish Lignite Mining," 112–14.

50. Ibid., 113–15; and Stefan Rahner, "Umsiedlung," in Ulrike Stottrop, ed., *Zeitraum Braunkohle* (1993), 107–14.

51. Frank Dickmann, *Umsiedlungsatlas des Rheinischen Braunkohlenreviers* (1996), 42–43.

52. Johannes Sigmond, "Landschaftswandel und Wasserhaushalt im rheinischen Braunkohlengebiet," in *Das rheinische Braunkohlengebiet: Eine Landschaft in Not!* (1953), 11–21. See also Hermann Josef Bauer, *Landschaftsökologische Untersuchungen im ausgekohlten rheinischen Braunkohlenrevier auf der Ville* (1963), 28–44.

53. Friedrich Schultz, *Deutscher Braunkohlen-Industrie-Verein e. V. 1885–1960* (n.d.), 53–55.

54. Matthias Schneider, *Wasserhaushalt und Wasserwirtschaft im Gebiete der Erftquellflüsse (Nordeifel)* (1953), 79–80.

55. Sigmond, "Landschaftswandel und Wasserhaushalt," 11–21. See also Bauer, *Landschaftsökologische Untersuchungen*, 28–44.

56. Franke and Frey, *Talsperren*, 386–401.

57. Arno Kleinebeckel, *Unternehmen Braunkohle: Geschichte eines Rohstoffs, eines Reviers, einer Industrie im Rheinland* (1986), 231.

58. Aldo Leopold, "Wilderness" (1935) in Susan L. Flader and J. Baird Callicott, eds., *The River of the Mother of God and Other Essays by Aldo Leopold* (1991), 226.

59. Ibid., 227–28.

60. Henry E. Lowood, "The Calculating Forester: Quantification, Cameral Science, and the Emergence of Scientific Forestry Management in Germany," in Tore Frängsmyr, J. L. Heilbron, and Robin E. Rider, eds., *The Quantifying Spirit in the 18th Century* (1990), 340–41.

61. Scott, *Seeing Like a State*, 21–22.

62. Cited by Rommelspacher, "Das natürliche Recht," 62.

63. Mumford, *Technics and Civilization*, 74.

Chapter 5. Sacrificing a River

1. Bundesministerium für Umwelt, Naturschutz und Reaktorsicherheit, "Rhein-Bericht: Bericht der Bundesregierung über die Verunreinigung des Rheins durch die Brandkatastrophe bei der Sandoz AG/Basel und weitere Chemieunfälle," *Umweltbrief*, no. 34 (February 1987); and Herbert Knöpp, "Der Rhein ein knappes Jahr nach 'Sandoz'," *Beiträge zur Rheinkunde*, no. 40 (1988): 5–19. See also Thomas Tittizer, Franz Schöll, and D. Hardt, "Die Lebensgemeinschaft der Rheinsohle," *Beiträge zur Rheinkunde*, no. 44 (1992): 47–54; and Thomas Tittizer, Franz Schöll, and Michael Schleuter, "Beitrag zur Struktur und Entwicklungsdynamik der Ben-

thalfauna des Rheins von Basel bis Düsseldorf in den Jahren 1986 und 1987," in Ragnar Kinzelbach and G. Friedrich, eds., *Biologie des Rheins* (1990), 293–323.

2. Sylvia Reckel, "Von 'Teufelsfarbe', 'Scharlachtüchern', 'Waidjunkern' und 'Schönfärben': Aufstieg und Fall der natürlichen Farben," in Arne Andersen and Gerd Spelsberg, eds., *Das blaue Wunder: Zur Geschichte der synthetischen Farben* (1990), 57–81; Jean Gimpel, *The Medieval Machine: The Industrial Revolution of the Middle Ages* (1976), 85–87; and Guillerme, *The Age of Water*, 99–100.

3. Kluge and Schramm, *Wassernöte*, 87–95.

4. Curt Weigelt, "Die Industrie und die preussische Ministerialverfügung von 20. Februar 1901," *Die Chemische Industrie* 24 (15 October 1901): 555.

5. See Gottfried Plumpe, *Die I. G. Farbenindustrie AG: Wirtschaft, Technik und Politik, 1904–1945* (1990), esp. 63–96 and 203–396. The closest equivalent to I. G. Farben was the British conglomerate, Imperial Chemical Industries, formed in 1926, in which Nobel Dynamite played a central role. In the United States, Du Pont was as close to a chemical-explosives conglomerate as trust-busting America would allow.

6. Aftalion, *A History of the International Chemical Industry*, 10–13; and Paul Hohenberg, *Chemicals in Western Europe, 1850–1914: An Economic Study of Technical Change* (1967), 23–25.

7. Aftalion, *A History of the International Chemical Industry*, 10–14; and Hohenberg, *Chemicals in Western Europe*, 23–24.

8. Aftalion, *A History of the International Chemical Industry*, 57–62; and Hohenberg, *Chemicals in Western Europe*, 24–25. Production statistics from Aftalion, 59.

9. Michael Dewhurst, "Fertiliser Progress 1841–1991: A Review of the Development of Mineral and Organic Fertilisers," in P. J. T. Morris, W. A. Campbell, and H. L. Roberts, eds., *Milestones in 150 Years of the Chemical Industry* (1991), 53–67; and Mirko Lamer, *The World Fertilizer Economy* (1957), 31–58.

10. J. McGrath, "Explosives," in Charles Singer, ed., *A History of Technology* (1954–84), vol. 5, 284–98; and Aftalion, *A History of the International Chemical Industry*, 55–57.

11. John Joseph Beer, *The Emergence of the German Dye Industry* (1959), 9–25; Hans Pohl, Ralf Schaumann, and Frauke Schönert-Röhlk, *Die chemische Industrie in den Rheinlanden während der industriellen Revolution* (1983), vol. 1, 24–28; Anthony S. Travis, "Synthetic Dyestuffs: Modern Colours for the Modern World," in Morris, Campbell, and Roberts, eds., *Milestones in 150 Years of the Chemical Industry*, 144–57; and Anthony S. Travis, *The Rainbow Makers: The Origins of the Synthetic Dyestuffs Industry in Western Europe* (1993). For a detailed list of available dyes and their properties as of 1910, see "Dyeing" in the eleventh edition of the *Encyclopaedia Britannica* (1910), vol. 8, 744–55.

12. Beer, *The Emergence of the German Dye Industry*, 99–100.

13. Ralf Henneking, *Chemische Industrie und Umwelt* (1994), 35–37; and Beer, *The Emergence of the German Dye Industry*, 53. See also Frauke Schönert-Röhlk, "Die räumliche Verteilung der chemischen Industrie im 19. Jahrhundert," in Hans

Pohl, ed., *Gewerbe- und Industrielandschaften vom Spätmittelalter bis ins 20. Jahrhundert* (1986), 417–55.

14. Ralf Schaumann, *Technik und technischer Fortschritt im Industrialisierungsprozeß: Dargestellt am Beispiel der Papier-, Zucker- und chemischen Industrie der nördlichen Rheinlande (1800–1875)* (1977), 479.

15. See Henneking, *Chemische Industrie und Umwelt*, table 1, 431–59.

16. Aftalion, *A History of the International Chemical Industry*, 39–42; and Beer, *The Emergence of the German Dye Industry*, 52–62.

17. Johann Peter Murmann and Ralph Landau, "On the Making of Competitive Advantage: The Development of the Chemical Industries in Britain and Germany Since 1850," in Ashish Arora, Ralph Landau, and Nathan Rosenberg, eds., *Chemicals and Long-Term Economic Growth: Insights from the Chemical Industry* (1998), 29. See also Aftalion, *A History of the International Chemical Industry*, 103–13; and Hohenberg, *Chemicals in Western Europe*, 40–42.

18. Aftalion, *A History of the International Chemical Industry*, 103–13; and Hohenberg, *Chemicals in Western Europe*, 40–42.

19. Lee Niedringhaus Davis, *The Corporate Alchemists: Profit Takers and Problem Makers in the Chemical Industry* (1984), 216–17.

20. Henneking, *Chemische Industrie und Umwelt*, 40–41; and F. Sherwood Taylor, *A History of Industrial Chemistry* (1957), 188–91.

21. Henneking, *Chemische Industrie und Umwelt*, 310–15.

22. Ibid., 40–41; and Taylor, *A History of Industrial Chemistry*, 188–91.

23. Cited by Davis, *The Corporate Alchemists*, 48.

24. Peter Brimblecombe, *The Big Smoke: A History of Air Pollution in London since Medieval Times* (1987), 136–38; and Davis, *The Corporate Alchemists*, 48–50. See also Robert Angus Smith, *Air and Rain: The Beginnings of Chemical Climatology* (1872).

25. Aftalion, *A History of the International Chemical Industry*, 14–15 and 57–62; Davis, *The Corporate Alchemists*, 50–54; and Henneking, *Chemische Industrie und Umwelt*, 52–53.

26. Henneking, *Chemische Industrie und Umwelt*, 57–59.

27. Karl Otto Henseling and Anselm Salinger, "'Eine Welt voll märchenhaften Reizes …,' Teerfarben: Keimzelle der modernen Chemieindustrie," in Andersen and Spelsberg, eds., *Das blaue Wunder*, 89–90; and Henneking, *Chemische Industrie und Umwelt*, 63–65.

28. Cited by Gerd Spelsberg, "'Im Fieber des Farbenrausches': Eine Siegesgeschichte," in Andersen and Spelsberg, eds., *Das blaue Wunder*, 46.

29. Tim Arnold, "'Ein leichter Geruch nach Fäulnis und Säure …,'" in Andersen and Spelsberg, eds., *Das blaue Wunder*, 154.

30. Peter Hüttenberger, "Umweltschutz vor dem Ersten Weltkrieg: Ein sozialer und bürokratischer Konflikt," in Hein Hoebink, ed., *Staat und Wirtschaft an Rhein und Ruhr, 1816–1991: 175 Jahre Regierungsbezirk Düsseldorf* (1992), 263–66.

31. Anthony S. Travis, "Poisoned Groundwater and Contaminated Soil: The

Tribulations and Trial of the First Major Manufacturer of Aniline Dyes in Basel," *Environmental History* 2 (July 1997): 343–65. See also Alfred Bürgin, *Geschichte des Geigy-Unternehmens von 1758 bis 1939* (1958), 118–19; and Markus Hämmerle, *Die Anfänge der Basler chemischen Industrie im Lichte von Arbeitsmedizin und Umweltschutz* (1979), 47–61.

32. Henseling and Salinger, "'Eine Welt voll märchenhaften Reizes...,'" 101–2.

33. Spelsberg, "'Im Fieber des Farbenrausches'," 24–25; and Henneking, *Chemische Industrie und Umwelt*, 61–62.

34. Egon Edwin Kisch, "Das giftige Königreich am Rhein," in Bodo Uhse and Gisela Kisch, eds., *Mein Leben für die Zeitung 1926–1947: Journalistische Texte 2* (1983), 48–49.

35. Spelsberg, "'Im Fieber des Farbenrausches'," 53.

36. Robert Henderson Clapperton, *Modern Paper-Making* (1952), 1–6.

37. Aftalion, *A History of the International Chemical Industry*, 81.

38. Sabine Schachtner, *Größer, schneller, mehr: Zur Geschichte der industriellen Papierproduktion und ihrer Entwicklung in Bergisch Gladbach* (1996), 37.

39. Hermann Schäfer, "Gewerbelandschaften: Elektro, Papier, Glas, Keramik," in Pohl, ed., *Gewerbe- und Industrielandschaften*, 470–71.

40. Schachtner, *Größer, schneller, mehr*, 77.

41. The fate of the Neffel is explained in detail in a letter dated 22 July 1913 from Leopold Peill (Düren) to the provincial authorities, in NRW/HSA, Regierung Köln, Nr. 8323.

42. Hermann Kellenbenz, *Die Zuckerwirtschaft im Kölner Raum von der Napoleonischen Zeit bis zur Reichsgründung* (1966), 35–52.

43. Günter Kufferath-Sieberin, *Die Zuckerindustrie der linksrheinischen Bördenlandschaft* (1955), 20–34.

44. A list of licensing controversies for Rhineland can be found in Henneking, *Chemische Industrie und Umwelt*, 431–60.

45. Ibid., 305.

46. For a fuller discussion of this legislation, see Ulrike Gilhaus, *"Schmerzenskinder der Industrie": Umweltverschmutzung, Umweltpolitik und sozialer Protest im Industriezeitalter in Westfalen, 1845–1914* (1995), 262–75.

47. "Wasserwirtschaftlicher Verband: Hauptversammlung, Berlin, 24. 2. 1912," *Zeitschrift für angewandte Chemie* 25 (26 April 1912): 835–36.

48. "Aktenvermerk über die Bereisung der Wupper von Leyersmühle bei Wipperfürth bis zur Wuppermündung (24–26 November 1913)," NRW/HSA, Regierung Köln, Nr. 8326.

49. Cited in Brüggemeier, "The Ruhr Basin," in Brimblecombe and Pfister, eds., *Silent Countdown*, 224.

50. *Solinger Tageblatt* (28 June 1922). Cited by Klaus-Georg Wey, *Umweltpolitik in Deutschland*, 94–95.

51. The various reports are available in NRW/HSA, Regierung Köln, Nrs. 8312–8313 and 8321–8327.

52. Johann Paul, "Die Opferstrecken wurden immer länger: Die Siegver-schmutzung im 19. und 20. Jahrhundert," in Wolfgang Isenberg, ed., *Historische Umweltforschung: Wissenschaftliche Neuorientierung—Aktuelle Fragestellungen* (Bergisch Gladbach: Thomas-Morus-Akademie Verlag, 1992), 57–71.

53. Mumford, *Technics and Civilization*, 212–67 (quotation, 255–56).

54. Schäfer, "Gewerbelandschaften," in Pohl, ed., *Gewerbe- und Industrieland-schaften*, 456–70.

55. Altmayer, "Vom Oberrhein zwischen Basel und Bodensee," *Der Schaffende Rhein: Neue Folge der Beiträge der Rheinfreunde*, no. 7 (1931): 4–22; Strack, "Der Hochrhein von Basel bis zum Bodensee," *Beiträge zur Rheinkunde*, no. 1 (1960): 27–37; and Ott and Herzig, "Elektrizitätsversorgung von Baden, Württemberg und Hohenzollern 1913/14," *Historischer Atlas von Baden-Württemberg*, vol. 11; and Calvis, "Die wasser- und energiewirtschaftliche Bedeutung des Rheins," 220–302.

56. Ulrich Linse, "'Der Raub des Rheingoldes': Das Wasserkraftwerk Laufen-burg," in Ulrich Linse, Reinhard Falter, Dieter Rucht, and Winfried Kretschmer, eds., *Von der Bittschrift zur Platzbesatzung: Konflikte um technische Großprojekte* (1988), 34.

57. Calvis, "Die wasser- und energiewirtschaftliche Bedeutung des Rheins," 27–31, 64–72, and 90–105.

58. See Kuster, "Die Geschichte der internationalen Rheinregulierung," Abb. 28; and Peter Matt, "Österreichische Kraftwerke im Rhein-Einzugsgebiet," in *Der Alpenrhein und seine Regulierung*, 352–54.

59. Pierre Miquel, *Histoire des canaux, fleuves et rivières de France* (1994), 233–62; and Calvis, "Die wasser- und energiewirtschaftliche Bedeutung des Rheins," 319–67.

60. Raymond G. Stokes, *Opting for Oil: The Political Economy of Technological Change in the West German Chemical Industry, 1945–1961* (1994), 2–3 and 137–38; and Aftalion, *A History of the International Chemical Industry*, 214–19.

61. Audrey M. Lambert, *The Making of the Dutch Landscape: An Historical Geography of the Netherlands* (1971), 347–52; and Egbert Wever, "The Port of Rotterdam," in Heineberg, de Lange, and Mayr, eds., *The Rhine Valley*, 1996), 11–18.

62. Horst Johannes Tümmers, *Der Rhein: Ein europäischer Fluß und seine Geschichte* (1994), 90.

63. Ibid., 332–35. For a brief overview of the antinuclear movement in Germany, see Saral Sarkar, *Green-Alternative Politics in West Germany* (1993–94), 97–146; and Raymond Dominick, *The Environmental Movement in Germany: Prophets and Pioneers, 1871–1971* (1992), 160–68 and 216–18. On French nuclear politics, see especially Gabrielle Hecht, *The Radiance of France: Nuclear Power and National Identity after World War II* (1998).

64. Dorothy Nelkin and Michael Pollak, *The Atom Besieged: Antinuclear Move-ments in France and Germany* (1982), 58–60.

65. Ibid., 60–64; and Linse, Falter, Rucht, and Kretschmer, eds., *Von der Bitt-*

schrift zur Platzbesetzung, 128–64.

66. Reichelt, *Laßt den Rhein leben!* 48–51.

67. Cited in Arne Andersen, "Pollution and the Chemical Industry: The Case of the German Dye Industry," in Ernst Homburg, Anthony S. Travis, and Harm G. Schröter, eds., *The Chemical Industry in Europe, 1850–1914: Industrial Growth, Pollution, and Professionalization* (1998), 196.

68. Wasserwirtschaftsverwaltung des Landes Nordrhein-Westfalen und der Wasser- und Schiffahrtsdirektion Duisburg, ed., *Die Verunreinigung des Rheins im Lande Nordrhein-Westfalen (Stand: Ende 1956): Maßnahmen zu ihrer Bekämpfung*, Anlage 12.

69. "Gutachtliche Äusserung über die Beschaffenheit des Rheinstromes auf der Strecke von Coblenz bis zur holländischen Grenze," Preussische Landesanstalt für Wasser-, Boden- und Lufthygiene (Tgb. Nr. 6925, Berlin-Dahlem, 25.8.1925), in NRW/HSA, Regierung Düsseldorf, Nr. 48469.

70. Thomas Tittizer and Falk Krebs, eds., *Ökosystemforschung: Der Rhein und seine Auen. Eine Bilanz* (1996), 42–48.

71. Ibid., 49–50.

72. Georg Ebeling, "Bericht über das Ergebnis der Rheinuntersuchungen in den Jahren 1935–1937," in NRW/HSA, Regierung Düsseldorf, Nr. 48472.

73. Tittizer and Krebs, eds., *Ökosystemforschung*, 51.

74. *Rhein-Zeitung* (27 June 1969 and 6 February 1970). Dutch researchers discovered that upstream from the Hoechst factory the Thiodan level stood at near-zero, while downstream the concentration stood at 4.3 micrograms per liter (with one microgram per liter being sufficient to kill off fish). Concentration levels dropped significantly once the Main mouthed into the Rhine, but traces (0.35 Hg/l) of Thiodan could be detected all the way to the Dutch-German border.

75. Davis, *The Corporate Alchemists*, 261–62.

76. Martin Forter and Horand Knaup, "Les dossiers noirs de la chimie—68 décharges sommeillent dans la Région des trois frontières," *Alsace* (20 December 1989); and Forter and Knaup, "Spiel mit Grenzen—Jahrelang verseuchten die Basler Chemiemultis die Umwelt am Oberrhein," *Die Zeit* (9 March 1990).

77. Aftalion, *A History of the International Chemical Industry*, 375–76; and Vollrath Hopp and Armin Beck, "Der Rhein," *Chemiker-Zeitung* 114 (July–August 1990): 229–43.

78. Thomas P. Hughes, *Networks of Power: Electrification in Western Society, 1880–1930* (1983), esp. 404–60.

Chapter 6. Biodiversity Lost

1. Mary Shelley, *Frankenstein; or, the Modern Prometheus* (1994 [1818]), 165–68 and 179.

2. Richard Boos and Rüdiger Krüpfganz, *Dampfboote und Kähne auf dem*

Rhein und seinen Nebenflüssen (1986), 5; and Wilhelm Treue, *150 Jahre Köln-Düsseldorfer: Die Geschichte der Personenschiffahrt auf dem Rhein* (1976), 7.

3. Hanns Dieter Schaake, *Der Fremdenverkehr in den linksrheinischen Kleinstädten zwischen Bingen und Koblenz* (1971), 8. See also Erdmann Gormsen, "The Cultural Landscape on the Upper Middle Rhine and in the Rheingau," in Heineberg, Lange, and Mayr, eds., *The Rhine Valley*, 153; and Walter Michels, *Unvergessene Dampfschiffahrt auf Rhein und Donau* (1967), 3.

4. Ulrike Pretzel, *Die Literaturform Reiseführer im 19. und 20. Jahrhundert: Untersuchungen am Beispiel des Rheins* (1995), 63.

5. Reichelt, *Laßt den Rhein leben!* 85–87.

6. T. H. Elkins and P. K. Marstrand, "Pollution of the Rhine and Its Tributaries," in Royal Institute of International Affairs, ed., *Regional Management of the Rhine: Papers of a Chatham House Study Group* (1975), 50–68; and L. J. Huizenga, "Suitable Measures against the Pollution of the Rhine by Chloride Discharges from the Alsatian Potash Mines," *Pure and Applied Chemistry* 29 (1972): 345–53.

7. The IAWR's findings were reported in P. Ph. Jansen, L. van Bendegom, J. van den Berg, M. de Vries, and A. Zanen, *Principles of River Engineering: The Non-Tidal Alluvial River* (1979), 26–28.

8. Der Rat von Sachverständigen für Umweltfragen, *Umweltprobleme des Rheins, 3. Sondergutachten März 1976* (1976), esp. 47–65.

9. *Rhein-Zeitung* (7 June 1977).

10. Rommelspacher, "Das natürliche Recht," in Brüggemeier and Rommelspacher, eds., *Besiegte Natur*, 42. See also Hans Grünfeld, *Creating Favorable Conditions for International Environmental Change through Knowledge and Negotiation* (1999), 39.

11. T. C. W. Blanning, *The French Revolution in Germany: Occupation and Resistance in the Rhineland, 1792–1802* (1983), 20.

12. Solmsdorf, "Schutzwürdigen Bereiche im Rheintal," *Beiträge zur Rheinkunde* 27 (1975): 28–31. See also Hartmut Solmsdorf, Wilhelm Lohmeyer, and Walter Mrass, *Ermittlung und Untersuchung der schutzwürdigen und naturnahen Bereiche entlang des Rheins (Schutzwürdige Bereiche im Rheintal)* (1975), 40 and 105. For an analysis of similar conditions on the Alpenrhein, see Thomas Kuster, "Die Geschichte der internationalen Rheinregulierung," 24.

13. Figures from *Baedeker's Rhine* (1985), 25–27. See also Siegfried Kolb, "Naturschutz und Landschaftspflege bei wasserbaulichen Maßnahmen, dargestellt am Beispiel des Oberrheins," *Beiträge zur Rheinkunde*, no. 38 (1986): 5–18.

14. Haebler, *Badische Geschichte*, 111.

15. Bruno Kremer, "Die Auengebiete des südlichen Oberrheins—Chance für die Natur oder verschwindende Naturlandschaft?" *Beiträge zur Rheinkunde*, no. 41 (1989): 56–67. The Taubergiessen was declared a nature protection area in 1979. See also Peter Obrdlik, Erika Schneider, and Renate Smukalla, "Zur Limnologie der Rastatter Rheinaue," in Kinzelbach and Friedrich, eds., *Biologie des Rheins*, 477–89, for a description of ecological conditions at the Rastatter Rheinaue (also

on the Upper Rhine), an area that came under protection in 1984. For a description of conditions on the Middle Rhine between the Ahr and Sieg tributaries, the site of several interlinked nature preserves, see Bruno Kremer, "Rheinaue und Auenvegetation im nördlichen Mittlerheingebiet," *Beiträge zur Rheinkunde*, no. 36 (1984): 38–49.

16. Emil Dister, Dieter Gomer, Petr Obrdlik, Peter Petermann, and Erika Schneider, "Water Management and Ecological Perspectives of the Upper Rhine's Floodplains," *Regulated Rivers: Research and Management* 5 (1990): 13.

17. *Baedeker's Rhine*, 26–27.

18. Tittizer and Krebs, eds., *Ökosystemforschung*, 198–99. See also Günther Niethammer, "Wandlungen in der Vogelwelt des Rheinlandes," *Rheinische Heimatpflege*, no. 3 (1964): 90–100; and Ulrich Wille, "Die Brutvögel zweier Baggerlöcher am unteren Niederrhein," *Rheinische Heimatpflege* 2 (1966): 139–53.

19. Tittizer and Krebs, eds., *Ökosystemforschung*, 199–204.

20. Ibid., 218–21.

21. Ibid., 195–97.

22. Ibid., 247–56; and Kinzelbach, ed., *Die Tierwelt des Rheins*, 36–40; more fully in Thomas Tittizer, Franz Schöll, and Maria Dommermuth, et al., "Zur Bestandsentwicklung des Zoobenthos des Rheins im Verlauf der letzten neun Jahrzehnte," *Wasser und Abwasser* 35 (1991): 125–66. See also Fred W. B. van den Brink, Gerard van der Velde, and Wobbe G. Cazemier, "The Faunistic Composition of the Freshwater Section of the River Rhine in The Netherlands: Present State and Changes since 1900," in Kinzelbach and Friedrich, eds., *Biologie des Rheins*, 191–216.

23. Tittizer and Krebs, eds., *Ökosystemforschung*, 247 and 251–53.

24. Kinzelbach, ed., *Die Tierwelt des Rheins einst und jetzt*, 37.

25. Tittizer and Krebs, eds., *Ökosystemforschung*, 194–95, 247, and 256.

26. Lelek and Buhse, *Fische des Rheins*, 34–35 and 186.

27. Ibid., 35–36.

28. Peter B. Moyle, Hiram W. Li, and Bruce A. Barton, "The Frankenstein Effect," in Richard H. Stroud, ed., *Fish Culture in Fisheries Management* (1986), 415–26.

29. Lelek and Buhse, *Fische des Rheins*, 30–32 and 43–44.

30. Ibid., 37–38.

31. Reichelt, *Laßt den Rhein leben!* 11–18 and 62–63; and Heinz Schurig, "Auswirkungen der Rheinregulierung auf die Fischerei," in *Der Alpenrhein und seine Regulierung*, 378–81. See also Annette Bernauer and Harald Jacoby, *Bodensee: Naturreichtum am Alpenrand* (1994), 28–33.

32. R. Demoll and H. N. Maier, eds., *Handbuch der Binnenfischerei Mitteleuropas*, vol. 5 (1926), 213.

33. Ibid., 211; and Lelek and Buhse, *Fische des Rheins*, 61–62.

34. Anna Schulte-Wülwer-Leidig, "Ecological Master Plan for the Rhine Catchment," in David M. Harper and Alastair J. D. Ferguson, eds., *The Ecological Basis*

for River Management (1995), 506.

35. Werner Böcking, "Der Niederrhein und seine Fischerei: Ein altes Handwerk ist zum Aussterben verurteilt," *Niederrheinisches Jahrbuch* 7 (1964): 103–4; Otto Drese, "Der Rhein und die Rheinfischerei," *Rheinische Heimatpflege* 12 (1975): 170–71; and Günter Jens and Ragnar Kinzelbach, "Der Lachs," *Mainzer Naturwissenschaftliches Archiv*, Beihefte 13 (1990): 58.

36. For the text of the treaty, see Bernd Rüster and Bruno Simma, eds., *International Protection of the Environment: Treaties and Related Documents*, vol. 9 (1975), 4724–29. See also Demoll and Maier, eds., *Handbuch der Binnenfischerei Mitteleuropas*, vol. 5, 210; and Otto Drese, "Der Rhein und die Rheinfischerei," *Rheinische Heimatpflege* 12 (1975): 162.

37. *Stenographische Berichte über die Verhandlungen des Reichstages*, 14 January 1897 (1871–1933), 4041.

38. Schulte-Wülwer-Leidig, "Ecological Master Plan," in Harper and Ferguson, eds., *The Ecological Basis for River Management*, 506. These figures correspond closely to those in Demoll and Maier, eds., *Handbuch der Binnenfischerei Mitteleuropas*, vol. 5, 208–9.

39. This letter (dated 30 September 1911) and similar ones are in NRW/HSA, Regierung Düsseldorf, Nr. 48468.

40. The Düsseldorf government report (dated 15 September 1911) is in NRW/HSA, Regierung Düsseldorf, Nr. 48468.

41. Letter reprinted in Boldt, ed., *Der Rhein*, 135.

42. "3. Rheinpolizeibezirk. Tgb.Nr.78/25. Gruppe Benrath, den 1.5.1925. Bericht," in NRW/HSA, Regierung Düsseldorf, Nr. 48469.

43. "Lachsfang in der preußischen Rheinprovinz im Jahre 1909/1910," in LHAK, Bestand 441/Nr. 25472.

44. *Jahresbericht des Rheinischen Fischerei-Vereins für 1908/09* (1909), 37. On the hatcheries, see "Lachsfischerei im Rheinstromgebiet," in LHAK, Bestand 441/Nr. 25472.

45. W. Fehlmann, *Die Ursachen des Rückganges der Lachsfischerei im Hochrhein* (1926), 5–6.

46. Ibid., 43–46. The first of the high dams, built at Rheinfelden in 1898, was low enough for salmon to surmount.

47. Jens and Kinzelbach, "Der Lachs," 60–61.

48. Günter Jens and Ragnar Kinzelbach, "Der Aal," *Mainzer Naturwissenschaftliches Archiv*, Beihefte. 13 (1990): 69–74; and Lelek and Buhse, *Fische des Rheins*, 160.

49. Tittizer and Krebs, eds., *Ökosystemforschung*, 272 and 286.

50. Kinzelbach, ed., *Die Tierwelt des Rheins einst und jetzt*, 37.

51. See Paul S. Giller and Björn Malmqvist, *The Biology of Streams and Rivers* (1998), 215–44.

Chapter 7. A River Restored?

1. Jörg Lange, "Robert Lauterborn (1869–1952)—Ein Leben am Rhein," *Lauterbornia*, no. 5 (September 1990): 1–25. For a representative sample of Lauterborn's works, see "Die Ergebnisse einer biologischen Probenuntersuchung des Rheins," *Arbeiten an der kaiserlichen Gesundheitsamt* 22 (1905): 630–52; "Die Vegetation des Oberrheins," *Verh. naturhistorisch-medizinischen Vereins Heidelberg* N.F. 10 (1910): 450–502; "Die geographische und biologische Gliederung des Rheinstromes," *Sitzungsbericht d. Heidelberger Akad. Wissensch. Math.-Nat.* 5–7 (1916–18): 1–61, 1–70, 1–87; and *Die erd- und naturkundliche Erforschung des Rheins und der Rheinlande vom Altertum bis zur Gegenwart*, vol. 1 of *Der Rhein: Naturgeschichte eines deutschen Stromes* (1934), 1–46.

2. Lange, "Robert Lauterborn," 9–10.

3. Ibid.

4. On ICPR's early history, see ICPR, *Bericht der Experten-Kommission über die physikalisch-chemische Untersuchung des Rheinwassers (1. Serie Juni 1953 bis Juni 1954)* (1956), 7–8; and Paul Sander, "Die Internationale Kommission zum Schutze des Rheins gegen Verunreinigung," *Beiträge zur Rheinkunde*, no. 22 (1970):14–18.

5. Treaty text in Rüster and Simma, eds., *International Protection of the Environment*, vol. 26, 1–10. See also Grünfeld, *Creating Favorable Conditions*, 24–27.

6. Grünfeld, *Creating Favorable Conditions*, 28–30; and Torsten and Tretter, "Der Rhein und Europa," in Boldt, ed., *Der Rhein*, 43.

7. Text in Rüster and Simma, eds., *International Protection of the Environment*, vol. 26, 440–49.

8. Ibid., vol. 4 (2d ser.), 7–20. For a detailed analysis of the cadmium problem, see William M. Stigliani and Stefan Anderberg, "Industrial Metabolism at the Regional Level: The Rhine Basin," in Robert U. Ayres and Udo E. Simonis, eds., *Industrial Metabolism: Restructuring for Sustainable Development* (1994), 119–62.

9. Text in Rüster and Simma, eds., *International Protection of the Environment*, vol. 4 (2d ser.), 124–27.

10. ICPR, *Aktionsprogramm "Rhein"* (1987), esp. 15–19 and 22–24. See also ICPR, *Aktionsprogramm "Rhein": Bestandaufnahme* (1989) and ICPR, *Aktionsprogramm "Rhein": Synthesebericht* (1989).

11. *Topic Rhine* (November 1991).

12. See especially the graphs in "Der Rhein—Ein Europäischer Fluß," *Umwelt: Monatszeitschrift des Bundesministers für Umwelt, Naturschutz und Reaktorsicherheit (Sonderausgabe)*, no. 9 (1988).

13. ICPR, *The Rhine: An Ecological Revival* (1993).

14. Hubert Hellmann, "Nitrat und Ammonium im Rhein—Konzentrationen, Frachten, Trendverhalten und Herkunft 1954–1988," *Zeitschrift für Wasser- und Abwasser-Forschung* 21 (1989): 212–22. See also *Topic Rhine* (September 1990).

15. Deutsche Kommission zur Reinhaltung des Rheins, *Rheinbericht 1990* (1991), 19–22.

16. Thomas Tittizer, Franz Schöll, and Maria Dommermuth, "The Development of the Macrozoobenthos in the River Rhine in Germany during the 20th Century," in J. A. van de Kraats, ed., *Rehabilitation of the River Rhine: Proceedings of the International Conference on Rehabilitation of the River, 15–19 March 1993, Arnhem, The Netherlands* (1994): 21–28. See also *Topic Rhine* (February 1991).

17. F. W. B. van den Brink, G. van der Velde, and A. bij de Vaate, "Ecological Aspects, Explosive Range Extension and Impact of a Mass Invader, *Corophium curvispinum* Sars, 1895 (Crustacea: Amphipoda) in the Lower Rhine (The Netherlands)," *Oecologia* 93 (1993): 224–32.

18. Tomás Brenner, Eberhard Hantge, and Ragnar Kinzelbach, "Zusammenfassung und Ergebnisse der Fachtagung am 21. November 1994 in Mainz," *Wie sauber ist der Rhein wirklich?—Biomonitoring* (1996), 143–48.

19. The origins of Salmon 2000 are covered in detail in two articles by Anne Schulte-Wülwer-Leidig: "Outline for the Ecological Master Plan for the Rhine," in Kraats, ed., *Rehabilitation of the River Rhine*, 273–80; and "Ecological Master Plan," in Harper and Ferguson, eds., *The Ecological Basis for River Management*, 505–14.

20. ICPR, *Tätigkeitsbericht 1992* (1993), 26–27; and A. W. De Haas and A. J. M. Smits, "Fish Passages in the Netherlands Rivers," in International Commission for the Hydrology of the Rhine Basin (CHR), *Ecological Rehabilitation of Floodplains* (1992), 89–93. See also Abraham bij de Vaate and André W. Breukelaar, "Sea Trout (*Salmo trutta*) Migration in the Rhine Delta, The Netherlands," and Elizabeth Hartgers, Tom Buijse, and Willem Dekker, "Situation for Migratory Fish in Lake IJsselmeer," both in ICPR, *2. Internationales Rhein-Symposium "Lachs 2000"* (1999), 80–84 and 85–91.

21. ICPR, *Tätigkeitsbericht 1992*, 26–27; and Beate Adam and Ulrich Schwevers, "Analyse des Fischwanderweges Lahn und Wiederansiedlung von Wanderfischen," in ICPR, *2. Internationales Rhein-Symposium "Lachs 2000,"* 127–34.

22. Stefan Jäger, "Lachs- und Meerforellenprogramm in Wupper, Dhünn und Ruhr," in ICPR, *2. Internationales Rhein-Symposium "Lachs 2000,"* 113–19.

23. ICPR, *Tätigkeitsbericht 1992*, 26–27; and Beate Adam and Ulrich Schwevers, "Analyse des Fischwanderweges Lahn und Wiederansiedlung von Wanderfischen," in ICPR, *2. Internationales Rhein-Symposium "Lachs 2000,"* 127–34.

24. Ulrich Schwevers, "Analyse des Fischwanderweges Main: der hessische Unterlauf," in ICPR, *2. Internationales Rhein-Symposium "Lachs 2000,"* 163–77.

25. Max Lauff, "Der Lachs im luxemburgischen Gewässersystem Sauer-Our," and Lothar Kroll, "Entwurf eines Entwicklungskonzeptes zur Verbesserung der ökologischen Funktionsfähigkeit der Mosel—unter besonderer Berücksichtigung des Fischwechsels," ibid., 145–49 and 150–60.

26. Martial Gerlinger, "Les plus grandes passes à poissons en Europe: Etat d'avancement des travaux de construction sur les chutes d'Iffezheim et de Gambsheim," ibid., 184–87.

27. P. Roche, "Habitat Availability and Carrying Capacity in the French Part

of the Rhine for Atlantic Salmon (Salar Salar L.)," in Kraats, ed., *Rehabilitation of the River Rhine*, 257–65; and Jean Jacques Klein and Jean François Luquet, "Situation au niveau du Rhin Supérieur: Actions 'Saumon 2000' en Alsace," in ICPR, *2. Internationales Rhein-Symposium "Lachs 2000,"* 188–94.

28. Roland Grimm, "'Lachs 2000' in Murg, Rench und Kinzig," in ICPR, *2. Internationales Rhein-Symposium "Lachs 2000,"* 195–99.

29. Peter Rey, Walter Hermann, Erich Staub, Claude Wisson, and Urs Zeller, "Lebensraum und Wiederansiedlung des Lachses in Basel," ibid., 204–10.

30. ICPR, *Ist der Rhein wieder ein Fluß für Lachse? "Lachs 2000"* (Koblenz: ICPR, 1999), 10–12.

31. "Pressemitteilung: Der Rhein—wieder ein Lachsfluß?," ibid., 297–302.

32. ICPR, *Ist der Rhein wieder ein Fluß für Lachse? "Lachs 2000,"* 43–48.

33. For a comprehensive overview of the Stork Plan, see Dick de Bruin, et al., *Ooievaar: De toekomst van het rivierengebied* (1987), 9–20. See also Eddy H. R. R. Lammens and Eric Marteijn, "Ecological Rehabilitation of Floodplains," CHR, *Ecological Rehabilitation*, 183–186.

34. H. Havinga, "Floodplain Restoration: A Challenge for River Engineering," in CHR, *Ecological Rehabilitation*, 95–103; and H. Duel, W. D. Denneman, and C. Kwakernaak, "Ecological Models for River Floodplain Rehabilitation," in Kraats, ed., *Rehabilitation of the River Rhine*, 383–86. See also Ministry of Transport, Public Works and Water Management, *Landscape Planning of the River Rhine in the Netherlands: Summary of the Main Report* (1996).

35. See especially Willem Overmars, "Ooibossen, een nieuw perspectief voor de uiterwaarden," and Frans Vera, "Fauna in het rivierenland," both in Bruin et al., *Ooievaar*, 35–46 and 47–58.

36. A. Siepe, "Regeneration of Floodplain Biotopes on the Upper Rhine—The 'Polder Altenheim' Case," and Jean-Paul Klein et al., "The Restoration of Former Channels in the Rhine Alluvial Forest: The Example of the Offendorf Nature Preserve (Alsace, France)," both in Kraats, ed., *Rehabilitation of the River Rhine*, 281–87 and 301–5.

37. *Topic Rhine* (October 1995 and September 1998).

38. ICPR, *Action Plan on Flood Defence* (1998), 11–13.

39. ICPR, *Rhein-Atlas: Ökologie und Hochwasserschutz* (Koblenz: ICPR, 1998).

40. ICPR, *Action Plan on Flood Defence*, 14–15.

41. Ibid., 20–21.

42. ICPR, *Bestandaufnahme der ökologisch wertvollen Gebiete am Rhein und erste Schritte auf dem Weg zum Biotopverbund* (1998), 9.

43. Ibid., 14–17.

44. Ibid., 49–55.

45. Cited by Günter Preuß, "Naturschutz," in Michael Geiger, Günter Preuß, Karl-Heinz Rothenberger, eds., *Der Rhein und die Pfälzische Rheinebene* (1991), 233.

46. Benno Wagner and Rudolf Ott, "Gewässerschutzmassnahmen im Rhein-

tal," in *Der Alpenrhein*, 348.

47. ICPR, *The Rhine: A River and Its Relations* (1998), 18–21.

48. H. Middelkoop and C. O. G. van Haselen, eds., *Twice a River: Rhine and Meuse in the Netherlands* (1999), 74 and 84.

Chapter 8. Conclusion

1. See Harper and Ferguson, eds., *The Ecological Basis for River Management*; P. H. Nienhuis, R. S. E. W. Leuven, and A. M. J. Ragas, eds., *New Concepts for Sustainable Management of River Basins* (1998); and B. Przedwojski, R. Blazejewski, and K. W. Pilarczyk, *River Training Techniques: Fundamentals, Design and Applications* (1995).

2. Bill Bryson, "Main-Danube Canal," *National Geographic* 182, no. 2 (August 1992): 3–31.

3. Hugo, *Le Rhin*, 152.

Bibliography

Unpublished Sources

Landeshauptarchiv Koblenz (LHAK), Koblenz, Germany
  Bestände 403, 418, 441
Nordrhein-Westfälisches Hauptstaatsarchiv (NRW/HSA), Düsseldorf, Germany
  Regierung Köln (Bestände 8312–8313, 8321–8327)
  Regierung Düsseldorf (Bestände 48468–48469, 48741–48742, 54009, 55904)

Published Sources

Aftalion, Fred. *A History of the International Chemical Industry.* Translated by Otto
   Theodor Benfey. Philadelphia: University of Pennsylvania Press, 1991.
Altmayer, Oberregierungsbaurat. "Vom Oberrhein zwischen Basel und Bodensee."
   Der Schaffende Rhein: Neue Folge der Beiträge der Rheinfreunde, no. 7 (1931):
   4–22.
Anderberg, Stefan, Sylvia Prieler, Krzysztof Olendrzynski, and Sander de Bruyn.
   *Old Sins: Industrial Metabolism, Heavy Metal Pollution, and Environmental
   Transition in Central Europe.* Tokyo: United Nations University Press, 2000.
Andersen, Arne. *Historische Technikfolgenabschätzung am Beispiel des Metallhütten-
   wesens und der Chemieindustrie, 1850–1933.* Stuttgart: Steiner, 1996.
Andersen, Arne, and Gerd Spelsberg, eds. *Das blaue Wunder: Zur Geschichte der
   synthetischen Farben.* Cologne: Volksblatt, 1990.
André, Fritz. *Bemerkungen über die Rectification des Oberrheins und Schilderung
   der furchtbaren Folgen, welche dieses Unternehmen für die Bewohner des Mittel-
   und Unterrheins nach sich ziehen wird.* Hanau: C. J. Edlerschen Buchhandlung,
   1828.
Ant, H. "Biologische Probleme der Verschmutzung und akuten Vergiftung von
   Fliessgewässern, unter besonderer Berücksichtigung der Rheinvergiftung im
   Sommer 1969." *Schriftenreihe für Landschaftspflege und Naturschutz* 4 (1969):
   97–126.
Arbeitsgemeinschaft der Länder zur Reinhaltung des Rheins. *Die Verunreinigung
   des Rheins und seiner wichtigsten Nebenflüsse in der Bundesrepublik Deutschland.*

Denkschrift (Stand 1971). Düsseldorf: Arbeitsgemeinschaft der Länder zur Reinhaltung des Rheins, 1972.

Arora, Ashish, Ralph Landau, and Nathan Rosenberg, eds. *Chemicals and Long-Term Economic Growth: Insights from the Chemical Industry*. New York: John Wiley & Sons, 1998.

Ayçoberry, Pierre, and Marc Ferro. *Une histoire du Rhin*. Paris: Editions Ramsay, 1981.

Ayres, Robert U., and Udo E. Simonis, eds. *Industrial Metabolism: Restructuring for Sustainable Development*. Tokyo: United Nations University Press, 1994.

Badischer Landwirtschaftlicher Hauptverband. *Steppe am Oberrhein? Der französische Rheinseitenkanal*. Freiburg im Breisgau: Badischer Landwirtschaftlicher Hauptverband, 1954.

Baedeker's Rhine. Stuttgart: Baedeker, 1985.

Banfield, Thomas C. *Industry of the Rhine* [1846–48]. New York: Augustus M. Kelley Publishers, 1969.

Barg, Friedbert, and Sandro Cambruzzi. *Schubeinheiten und Koppelverbände*. Herford: Koehler, 1991.

Bärthel, Ernst-Volker. *Der Stadtwald Breisach: 700 Jahre Waldgeschichte in der Aue des Oberrheins*. Stuttgart: Schriftenreihe der Landesfortverwaltung Baden-Württemberg, 1965.

Bauer, Hermann Josef. *Landschaftsökologische Untersuchungen im ausgekohlten rheinischen Braunkohlenrevier auf der Ville*. Arbeiten zur Rheinischen Landeskunde, no. 19. Bonn: Ferd. Dümmlers, 1963.

Bax, Jack, and J. Breadvelt. "Die Mündung des Rheins: Rotterdam und Europoort." *Welt am Oberrhein* 10, no. 4 (1970): 198–206.

Beer, John Joseph. *The Emergence of the German Dye Industry*. Urbana: University of Illinois Press, 1959.

Bergh, F. van den. *Die Felsen-Sprengungen im Rhein bei Bingen zur Erweiterung des Thalweges im Binger-Loche*. Koblenz: Karl Bädeker, 1834.

Bernauer, Annette, and Harald Jacoby. *Bodensee: Naturreichtum am Alpenrand*. Überlingen: Naturerbe Verlag, 1994.

Bernhardt, Christoph. "Zeitgenössische Kontroversen über die Umweltfolgen der Oberrheinkorrektion im 19. Jahrhundert." In *Zeitschrift für die Geschichte des Oberrheins*, 293–319. Stuttgart: W. Kohlhammer, 1998.

Betz, Helmut. *Die großen Motorschlepper und die Entwicklung der Schubschiffahrt auf dem Rhein*. Vol. 4 of *Historisches vom Strom*. Duisburg: Krüpfganz, 1988.

Beyerhaus, E. *Der Rhein von Strassburg bis zur Holländischen Grenze in technischer und wirtschaftlicher Beziehung*. Koblenz: Königliche Rheinstrombauverwaltung, 1902.

Biemond, Cornelius. "Rhine River Pollution Studies." *Journal of American Water Works Association* 63, no. 1 (1971): 36–40.

Blanning, T. C. W. *The French Revolution in Germany: Occupation and Resistance in the Rhineland, 1792–1802*. Oxford: Clarendon, 1983.

Böcking, Werner. "Der Niederrhein und seine Fischerei: Ein altes Handwerk ist zum Aussterben verurteilt." *Niederrheinisches Jahrbuch* 7 (1964): 101–7.

———. "Das 'Flugnetz' und seine Beanstandungen im Lachsvertrag 1891." *Der Niederrhein* 40, no. 2 (April 1973): 74–76.

———. "Störe waren einst wertvolle 'Beifänge'." *Der Niederrhein* 41, no. 3 (July 1974): 114–15.

———. "Bedeutende Salmfänge des Ober- und Mittelrheins." *Beiträge zur Rheinkunde*, no. 31 (1979): 52–60.

———. *Die Geschichte der Rheinschiffahrt: Schiffe auf dem Rhein in drei Jahrtausenden.* Moers: Aug. Steiger, 1980.

Boldt, Hans, ed. *Der Rhein: Mythos und Realität eines europäischen Stromes.* Cologne: Rheinland-Verlag, 1988.

Boos, Richard, and Rüdiger Krüpfganz. *Dampfboote und Kähne auf dem Rhein und seinen Nebenflüssen.* Vol. 3 of *Historisches vom Strom.* Duisburg: Krüpfganz, 1986.

Brenner, Tomás, Eberhard Hantge, and Ragnar Kinzelbach. "Zusammenfassung und Ergebnisse der Fachtagung am 21. November 1994 in Mainz." In *Wie sauber ist der Rhein wirklich?—Biomonitoring*, 143–48. Petersberg: Advanced Biology, 1996.

Brimblecombe, Peter. *The Big Smoke: A History of Air Pollution in London since Medieval Times.* London: Methuen, 1987.

Brimblecombe, Peter, and Christian Pfister, eds. *The Silent Countdown: Essays in European Environmental History.* Berlin: Springer, 1990.

Brink, F. W. B. van den, G. van der Velde, and A. bij de Vaate. "Ecological Aspects, Explosive Range Extension and Impact of a Mass Invader, *Corophium curvispinum* Sars, 1895 (Crustacea: Amphipoda) in the Lower Rhine (The Netherlands)." *Oecologia* 93 (1993): 224–32.

Brüggemeier, Franz-Josef. "A Nature Fit for Industry: The Environmental History of the Ruhr Basin, 1840–1990." *Environmental History Review* 18, no. 1 (Spring 1994): 35–54.

Brüggemeier, Franz-Josef, and Thomas Rommelspacher, eds. *Besiegte Natur: Geschichte der Umwelt im 19. und 20. Jahrhundert.* Munich: C. H. Beck, 1987.

———. *Blauer Himmel über der Ruhr: Geschichte der Umwelt im Ruhrgebiet, 1840–1990.* Essen: Klartext, 1992.

Bruin, Dick de, et al. *Ooievaar: De toekomst van het rivierengebied.* Arnhem: Stichting Gelderse Milieufederatie, 1987.

Bryson, Bill. "Main-Danube Canal." *National Geographic* 182, no. 2 (August 1992): 3–31.

Bürgin, Alfred. *Geschichte des Geigy-Unternehmens von 1758 bis 1939.* Basel: Geigy, 1958.

Büschenfeld, Jürgen. *Flüsse und Kloaken: Umweltfragen im Zeitalter der Industrialisierung (1870–1918).* Stuttgart: Klett-Cotta, 1997.

Bundesanstalt für Gewässerkunde. "Faunistische Erhebungen an der Rheinsohle zur Feststellung und Bewertung der Schädigung der Benthosbiozönose durch den Brand bei der Fa. Sandoz in Basel." In *Forschungsbericht*. Koblenz: Bundesanstalt für Gewässerkunde, 1987.

———. *Das Hochwasser 1993/94 im Rheingebiet*. Bfg-Nr. 0833. Koblenz: Bundesanstalt für Gewässerkunde, 1994.

Bundesministerium für Umwelt, Naturschutz und Reaktorsicherheit. "Rhein-Bericht: Bericht der Bundesregierung über die Verunreinigung des Rheins durch die Brandkatastrophe bei der Sandoz AG/Basel und weitere Chemieunfälle." *Umweltbrief*, no. 34 (February 1987).

———. "Der Rhein—Ein Europäischer Fluß: Beispiel und Herausforderung für den Gewässerschutz." *Umwelt (Sonderausgabe)*, no. 9 (1988).

Bundestagausschusses für Umwelt, Naturschutz und Reaktorsicherheit. *Zur Sache: Themen parlamentarischer Beratung—Schutz der Nordsee*. Bonn: Bundestagausschusses für Umwelt, Naturschutz und Reaktorsicherheit, 1987.

Buschmann, Walter, ed. *Koks, Gas, Kohlechemie: Geschichte und Gegenständliche Überlieferung der Kohleveredlung*. Essen: Klartext, 1993.

Bussche-Haddenhausen, Dethard Freiherr von dem. "Einiges über die Geschichte und Tätigkeit der Zentralkommission für die Rheinschiffahrt." *Beiträge zur Rheinkunde*, no. 19 (1968), 13–29.

Byron, Lord. *Childe Harold's Pilgrimage and Other Romantic Poems*. Edited by Samuel C. Chew (1819). New York: Odyssey, 1936.

Caesar, Julius. *The Gallic War*. The Loeb Classical Library. Cambridge: Harvard University Press, 1963.

Calvis, Herbert. "Die wasser- und energiewirtschaftliche Bedeutung des Rheins von seinen Quellen bis zum Eintritt ins Rheinische Schiefergebirge." Ph.D. dissertation, Cologne University, 1981.

Carling, P. A., and G. E. Petts, eds. *Lowland Floodplain Rivers: Geomorphological Perspectives*. Chichester: John Wiley & Sons, 1992.

Cassinone, Heinrich, and Karl Spieß. *Johann Gottfried Tulla: Sein Leben und Wirken*. Karlsruhe: C. F. Müller, 1929.

Centralbureau für Meteorologie und Hydrographie im Großherzogthum Baden. *Der Rheinstrom und seine wichtigsten Nebenflüsse*. Berlin: Ernst & Korn, 1889.

Chamberlain, J. P. "The Regime of the International Rivers: Danube and Rhine." Ph.D. dissertation, Columbia University, 1923.

Clapp, Edwin J. *The Navigable Rhine*. Boston: Houghton Mifflin, 1911.

Clapperton, Robert Henderson. *Modern Paper-Making*. Oxford: Basil Blackwell, 1952.

Coleridge, Samuel Taylor. *The Complete Poetical Works of Samuel Taylor Coleridge*. Vol. 1. Edited by Ernest Hartley Coleridge (1912). Oxford: Clarendon Press, 1957.

Commission Centrale pour la Navigation du Rhin. *Rapport Annuel de la Commission Centrale pour la Navigation du Rhin*. Strasbourg: Commission Centrale pour la Navigation du Rhin, 1985.

Cornelsen Verlag. *The Rhine: The World's Greatest Commercial River*. Cheltenham, England: European Schoolbooks, 1994.

Cosgrove, Denis, and Geoff Petts. *Water, Engineering and Landscape: Water Control and Landscape Transformation in the Modern Period*. London: Belhaven, 1990.

Cowx, Ian G., and Robin L. Welcomme, eds. *Rehabilitation of Rivers for Fish: A Study Undertaken by the European Inland Fisheries Advisory Commission of FAO*. Oxford: Fishing News Books, 1998.

Davis, Lee Niedringhaus. *The Corporate Alchemists: Profit Takers and Problem Makers in the Chemical Industry*. New York: William Morrow, 1984.

Dehn, Arne, Martin Desch, Lars Deckert, Jens Gallenbacher, Rolf Hartmann. *Auch Umwelt hat Geschichte: Eine Fabrik von 1873 bis heute*. Bad Homburg: Verlag Ausbildung und Wissen, 1987.

De la Suisse à la Mer: Edition Spéciale de la Revue Schweizerland, September 1920.

Demangeon, Albert, and Lucien Febvre. *Le Rhin: Problèmes d'histoire et d'économie*. Paris: Armand Colin, 1935.

Demoll, R., and H. N. Maier, eds. *Handbuch der Binnenfischerei Mitteleuropas*. Vol. 5. Stuttgart: E. Schweizerbart'sche Verlagsbuchhandlung, 1926.

Der Alpenrhein und seine Regulierung: Internationale Rheinregulierung, 1892–1992. Rorschach: Internationale Rheinregulierung, 1992.

Der Hafenkurier (Sonderausgabe), February 1972.

Der Rat von Sachverständigen für Umweltfragen. *Umweltprobleme des Rheins, 3 Sondergutachten März 1976*. Stuttgart: W. Kohlhammer, 1976.

"Der Rhein, ein europäisches Umweltproblem." In *Aktuelle JRO-Landkarte*. Munich: JRO-Verlag, 1985.

Der Ruhr—Schiffahrtsweg. Duisburg: Staatliches Amt für Wasser- und Abfallwirtschaft Herten, 1990.

Deutsche Forschungsgemeinschaft. *Hydrologischer Atlas der Bundesrepublik Deutschland*. Boppard: Harald Boldt, 1979.

Deutsche Kommission zur Reinhaltung des Rheins. *Rheinbericht 1990*. Düsseldorf: Deutsche Kommission zur Reinhaltung des Rheins, 1991.

Dickmann, Frank. *Umsiedlungsatlas des Rheinischen Braunkohlenreviers*. Cologne: Rheinland-Verlag, 1996.

Dieterlen, J. "Kembs: Premier Échelon du Grand Canal d'Alsace." *La Navigation du Rhin* 10 (November 1932) 405–69.

Dister, Emil, Dieter Gomer, Petr Obrdlik, Peter Petermann, and Erika Schneider. "Water Management and Ecological Perspectives of the Upper Rhine's Floodplains." *Regulated Rivers: Research and Management* 5 (1990): 1–15.

Dominick, Raymond H. *The Environmental Movement in Germany: Prophets and Pioneers, 1871–1971*. Bloomington: Indiana University Press, 1992.

Doyle, Richard. *The Foreign Tour of Messrs. Brown, Jones, and Robinson*. London: Bradbury & Evans, 1855.

Drese, Otto. "Der Rhein und die Rheinfischerei." *Rheinische Heimatpflege* 12

(1975): 161–74 and 241–55.

Dumas, Alexandre. *Excursions sur les bords du Rhin*. New York: American Book Company, 1905.

Dyke, George Mansford. "Castles on the Rhine: Romanticism, Nationalism, and Place in Nineteenth-Century Germany." Master's thesis, UCLA, 1994.

Ebeling, Dietrich. *Der Holländerholzhandel in den Rheinlanden: zu den Handelsbeziehungen zwischen den Niederlanden und dem westlichen Deutschland im 17. und 18. Jahrhundert*. Stuttgart: Steiner, 1992.

Eckert, Christian. *Rheinschiffahrt im XIX. Jahrhundert*. Leipzig: Verlag von Duncker & Humblot, 1900.

Eckoldt, Martin, ed. *Flüsse und Kanäle: Die Geschichte der deutschen Wasserstrassen*. Hamburg: DSV-Verlag, 1998.

———. "Johann Gottfried Tulla—Zu seinem 200. Geburtstag." *Beiträge zur Rheinkunde*, no. 22 (1970): 19–22.

Eliot, T. S. *The Dry Salvages*. London: Faber and Faber, 1941.

Elkins, T. H., and E. M. Yates. "The Neuwied Basin." *Geography* 45 (1960): 39–51.

Fehlmann, W. *Die Ursachen des Rückganges der Lachsfischerei im Hochrhein*. Beilage zum Jahresbericht der Kantonsschule Schaffhausen. Schaffhausen: Meier & Cie, 1926.

Felkel, Karl. "Strombau-Geschichte der Binger-Loch-Strecke des Rheins." *Beiträge zur Rheinkunde*, no. 12 (1961): 26–44.

———. "Das Problem der Sohlenstabilisierung des Oberrheins und die Naturversuche mit Geschiebezugabe." *Beiträge zur Rheinkunde*, no. 33 (1981): 20–35.

Fenner, Max, ed. *Die Binnenschiffahrt und Wasserkraftnutzung der Schweiz*. Zürich: Polygraphisches Institut, 1926.

Fenzl, Manfred. *Der Rhein: Schaffhausen-Nordsee und zum IJsselmeer*. Hamburg: Maritim, 1994.

Flader, Susan L., and J. Baird Callicott, eds. *The River of the Mother of God and Other Essays by Aldo Leopold*. Madison: University of Wisconsin Press, 1991.

Flood Control Measures in the Rhine Basin: Country Monograph Submitted by the Government of the Federal Republic of Germany. Seminar on Co-Operation in the Field of Transboundary Waters, Düsseldorf, 15–19 October 1984. Brussels: United Nations, Economic Commission of Europe, Committee on Water Problems, 1984.

Föhl, Axel, and Manfred Hamm. *Die Industriegeschichte des Wassers*. Düsseldorf: VDI-Verlag, 1986.

Forster, George. *Ansichten vom Niederrhein von Brabant, Flandern, Holland, England und Frankreich, im April, Mai und Junius 1790*. Frankfurt: Insel, 1969.

Forter, Martin, and Horand Knaup. "Les dossiers noirs de la chimie—68 décharges sommeillent dans la Région des trois frontières." *Alsace*, 20 December 1989.

———. "Spiel mit Grenzen—Jahrelang verseuchten die Basler Chemiemultis die Umwelt am Oberrhein." *Die Zeit*, 9 March 1990.

Frängsmyr, Tore, J. L. Heilbron, and Robin E. Rider, eds. *The Quantifying Spirit in the 18th Century*. Berkeley: University of California, 1990.

Franke, Peter, and Wolfgang Frey. *Talsperren in der Bundesrepublik Deutschland*. Berlin: Nationales Komitee für Große Talsperren in der Bundesrepublik, 1987.

Froehlich-Schmitt, Barbara. *Salmon 2000*. Koblenz: ICPR, 1994.

Gaay, A. C. de. *The Canalization of the Lower Rhine*. N.p.: Rijkswaterstaat Communications, 1970.

Garritsen, Ton, Guido Vonk, and Kees de Vries, eds. *Visions for the Rhine*. Lelystad: National Institute for Inland Water Management and Waste Water Treatment (RIZA), 2000.

Geiger, Michael, Günter Preuß, and Karl-Heinz Rothenberger, eds. *Der Rhein und die Pfälzische Rheinebene*. Landau i. d. Pfalz: Verlag Pfälzische Landeskunde, 1991.

Geißen, Hans-Peter. "Neue Gesellschaften im Rhein." *Kommune: Forum für Politik, Ökonomie und Kultur* (1991), 46–48.

Gerken, B. *Auen—verborgene Lebensadern der Natur*. Freiburg im Breisgau: Rombach, 1988.

Gerlach, Renate. *Flußdynamik des Mains unter dem Einfluß des Menschen seit dem Spätmittelalter*. Forschungen zur Deutschen Landeskunde. Trier: Zentralausschuss für Deutsche Landeskunde, 1990.

Gilhaus, Ulrike. *"Schmerzenskinder der Industrie": Umweltverschmutzung, Umweltpolitik und sozialer Protest im Industriezeitalter in Westfalen, 1845–1914*. Paderborn: Ferdinand Schöningh, 1995.

Giller, Paul S., and Björn Malmqvist. *The Biology of Streams and Rivers*. Oxford: Oxford University Press, 1998.

Gimpel, Jean. *The Medieval Machine: The Industrial Revolution of the Middle Ages*. New York: Holt, Rinehart and Winston, 1976.

Goubert, Jean-Pierre. *The Conquest of Water: The Advent of Health in the Industrial Age*. Translated by Andrew Wilson. Princeton: Princeton University, 1989.

Governmental Institute for Waste Water Treatment. *The Water Quality of the River Rhine in the Netherlands over the Period 1970–1981*. Ministry of Transport and Public Works (Rijkswaterstaat), 1983.

Greenpeace. *Der Rhein—kein Vorbild für die Elbe: Greenpeace Studie*. Hamburg: Greenpeace, 1991.

Gregory, K. J., ed. *River Channel Changes*. Chichester: John Wiley, 1977.

Grünfeld, Hans. *Creating Favorable Conditions for International Environmental Change through Knowledge and Negotiation*. Delft: Delft University Press, 1999.

Guillerme, André E. *The Age of Water: The Urban Environment in the North of France, A.D. 300–1800*. College Station: Texas A & M University, 1983.

Haebler, Rolf Gustav. *Badische Geschichte*. Karlsruhe: G. Braun, 1951.

Häringer, Georg. "Die Verunreinigung des Rheins." *Beiträge zur Rheinkunde*, no. 23 (1971): 25–33.

Hahn, Helmut, and Wolfgang Zorn, eds. *Historische Wirtschaftskarte der Rhein-*

lande um 1820. Rheinisches Archiv, vol. 87. Bonn: Ludwig Röhrscheid, 1973.

Hardenberg, Theo. "Der Drachenfels—Seine 'Conservation vermittelst Expropriation'." *Rheinische Heimatpflege*, no. 4 (1968): 274–310.

Hardin, Garrett. "The Tragedy of the Commons." *Science* 162 (1968): 1243–48.

Harper, David M., and Alastair J. D. Ferguson, eds. *The Ecological Basis for River Management*. Chichester: John Wiley & Sons, 1995.

Haupt, Thea. *Das Buch vom großen Strom*. Reutlingen: Ensslin & Laiblin, 1961.

Hecht, Gabrielle. *The Radiance of France: Nuclear Power and National Identity after World War II*. Cambridge: MIT Press, 1998.

Heidegger, Martin. *The Question Concerning Technology and Other Essays*. Translated by William Lovitt. New York: Garland Publishing, 1977.

Heimbrecht, Jörg. *Rheinalarm: Die genehmigte Vergiftung*. Cologne: Pahl-Rugenstein, 1987.

Heineberg, Heinz, Norbert de Lange, and Alois Mayr, eds. *The Rhine Valley— Urban, Harbour and Industrial Development and Environmental Problems: A Regional Guide Dedicated to the 28th International Geographical Congress, The Hague 1996*. Beiträge zur Regionalen Geographie, vol. 41. Leipzig: Institut für Länderkunde, 1996.

Heinrichsbauer, August. *Die Wasserwirtschaft im rheinisch-westfälischen Industriegebiet*. Essen: Boeckling & Müller, 1936.

———. *Industrielle Siedlung im Ruhrgebiet in Vergangenheit, Gegenwart und Zukunft*. Essen: Glückauf, 1936.

Hellmann, Hubert. "Nitrat und Ammonium im Rhein—Konzentrationen, Frachten, Trendverhalten und Herkunft 1954–1988." *Zeitschrift für Wasser- und Abwasser-Forschung* 21 (1989): 212–22.

———. "Die Verschmutzung des Rheins in den letzten Jahrzehnten." *Beiträge zur Rheinkunde*, no. 45 (1993): 49–60.

Helmer, Franz. "Die Ruhr: Eine kulturhistorische, wasser- und energiewirtschaftliche Betrachtung." *Beiträge zur Rheinkunde*, no. 41 (1989): 35–55.

Henneking, Ralf. *Chemische Industrie und Umwelt: Konflikte um Umweltbelastungen durch die chemische Industrie am Beispiel der schwerchemischen, Farben- und Düngemittelindustrie der Rheinprovinz (ca. 1800–1914)*. Stuttgart: Steiner, 1994.

Henseling, Karl Otto. *Ein Planet wird vergiftet: Der Siegeszug der Chemie: Geschichte einer Fehlentwicklung*. Reinbek bei Hamburg: Rowohlt, 1992.

Herendeen, Wyman H. *From Landscape to Literature: The River and the Myth of Geography*. Pittsburgh: Duquesne University Press, 1986.

Herman, Zvi. *The River and the Grain*. Translated by Asher Goldstin. New York: Herzl Press, 1988.

Hermand, Jost. *Grüne Utopien in Deutschland: Zur Geschichte des Ökologischen Bewußtseins*. Frankfurt: Fischer Taschenbuch, 1991.

Hessische Landesanstalt für Umwelt. *Verbesserung der Umweltverhältnisse auf dem Rhein*. Wiesbaden: Hessischer Minister für Landesentwicklung, Umwelt, Landwirtschaft und Forsten, 1977–78.

Hinrich, Helmut. "Die Schwebstofffracht des Rheins." *Beiträge zur Rheinkunde*, no. 31 (1979): 61–66.

Hoebink, Hein, ed. *Staat und Wirtschaft an Rhein und Ruhr, 1816–1991: 175 Jahre Regierungsbezirk Düsseldorf.* Essen: Klartext, 1992.

Hoek, P. P. C. "Propagation and Protection of the Rhine Salmon." *Bulletin of the U.S. Bureau of Fisheries* 28 (1908): 819–29.

Hölderlin, Friedrich. *Hymns and Fragments.* Translated by Richard Sieburth. Princeton: Princeton University Press, 1984.

Hölscher, Georg. *Das Buch vom Rhein: Eine Schilderung des Rheinstromes und seiner Ufer von den Quellen bis zum Meere unter besonderer Berücksichtigung seiner 2000jährigen Geschichte.* Cologne: Verlag von Hoursch & Bechstedt, 1927.

Hoffmann, Godehard von. *Rheinische Romanik im 19. Jahrhundert: Denkmalpflege in der preussischen Rheinprovinz.* Cologne: J. P. Bachem, 1995.

Hohenberg, Paul. *Chemicals in Western Europe, 1850–1914: An Economic Study of Technical Change.* Chicago: Rand McNally, 1967.

Homburg, Ernst, Anthony S. Travis, and Harm G. Schröter. *The Chemical Industry in Europe, 1850–1914: Industrial Growth, Pollution, and Professionalization.* Dordrecht: Kluver, 1998.

Honnef, Klaus, Klaus Weschenfelder, and Irene Haberland, eds. *Vom Zauber des Rheins ergriffen ... Zur Entdeckung der Rheinlandschaft.* Munich: Klinkhardt & Biermann, 1992.

Honsell, Max. *Die Hochwasser-Katastrophen am Rhein im November und Dezember 1882.* Berlin: Ernst & Korn, 1883.

———. *Die Korrektion des Oberrheins von der Schweizer Grenze unterhalb Basel bis zur Großherzogthum Hessischen Grenze unterhalb Mannheim.* Karlsruhe: Druck der G. Braun'schen Hofbuchdruckerei, 1885.

Hopp, Vollrath, and Armin Beck. "Der Rhein—größter Standort der chemischen Industrie in der Welt." *Chemiker-Zeitung* 114 (July–August 1990): 229–43.

Hoppe, Christine. *Die großen Flußverlagerungen des Niederrheins in den letzten zweitausend Jahren und ihre Auswirkungen auf Lage und Entwicklung der Siedlungen.* Forschungen zur deutschen Landeskunde. Bad Godesberg: Bundesanstalt für Landeskunde und Raumordnung, 1970.

Hübner, Paul. *Der Rhein: Von den Quellen bis zu den Mündungen.* Frankfurt: Societäts-Verlag, 1974.

Hueting, R. "Some Economic Aspects of Pollution of the Rhine." In *Rhine Pollution: Legal, Economic and Technical Aspects.* Zwolle, Netherlands: Tjeenk Willink, 1978.

Hughes, Thomas P. *Networks of Power: Electrification in Western Society, 1880–1930.* Baltimore: Johns Hopkins University Press, 1983.

Hugo, Victor. *Le Rhin: Lettres à un ami, 1838–1839.* Strasbourg: Bueb et Reumaux, 1980.

Huisman, Peter. "Internationale Zusammenarbeit zur Bekämpfung der Verunreinigung des Rheins." *Beiträge zur Rheinkunde*, no. 32 (1980): 42–50.

Huizenga, L. J. "Suitable Measures against the Pollution of the Rhine by Chloride Discharges from the Alsatian Potash Mines." *Pure and Applied Chemistry* 29 (1972): 345–53.

International Commission for the Hydrology of the Rhine Basin (CHR). *Das Rheingebiet—Hydrologische Monographie.* The Hague: CHR, 1977.

———. *Die Hochwasser an Rhein und Mosel im April und Mai 1983.* Report No. II-3 of CHR/KHR. Lelystad, Netherlands: CHR, 1989.

———. *Das Hochwasser 1988 im Rheingebiet.* Report No. I-9 of the CHR/KHR. Lelystad, Netherlands: CHR, 1990.

———. *Ecological Rehabilitation of Floodplains: Contributions to the European Workshop, Arnhem, The Netherlands, 22–24 September 1992.* Report No. II-6 of the CHR/KHR. The Haag: CIP-Gegevens Koninklijke Bibliotheek, 1992.

———. *Der Rhein unter der Einwirkung des Menschen—Ausbau, Schiffahrt, Wasserwirtschaft.* Lelystad: CHR, 1993.

International Commission for the Protection of the Rhine (ICPR). *Bericht der Experten-Kommission über die physikalisch-chemische Untersuchung des Rheinwassers (1. Serie Juni 1953 bis Juni 1954).* Basel: Birkhäuser, 1956.

———. *Aktionsprogramm "Rhein."* Strasbourg: ICPR, 1987.

———. *Aktionsprogramm "Rhein": Bestandaufnahme der Einleitungen prioritärer Stoffe 1985 und Vorausschau über die bis 1995 erzielbaren Verringerungen der Einleitungen.* Brussels: ICPR, 1989.

———. *Aktionsprogramm "Rhein": Synthesebericht über die z. Z. laufenden und bereits geplanten Maßnahmen zur Verbesserung des Ökosystems 'Rhein' inkl. seiner Nebengewässer.* Brussels: ICPR, 1989.

———. *Zahlentafeln der physikalisch-chemischen Untersuchungen des Rheinwassers und des Schwebstoffs 1990.* Koblenz: ICPR, 1990.

———. *The Rhine: An Ecological Revival.* Koblenz: ICPR, 1993.

———. *Tätigkeitsbericht 1992.* Koblenz: ICPR, 1993.

———. *Statusbericht Rhein 1997.* Koblenz: ICPR, 1997.

———. *Action Plan on Flood Defence.* Koblenz: ICPR, 1998.

———. *Bestandaufnahme der ökologisch wertvollen Gebiete am Rhein und erste Schritte auf dem Weg zum Biotopverband.* Koblenz: ICPR, 1998.

———. *Rhein-Atlas: Ökologie und Hochwasserschutz.* Koblenz: ICPR, 1998.

———. *The Rhine: A River and Its Relations.* Koblenz: ICPR, 1998.

———. *2. Internationales Rhein-Symposium "Lachs 2000."* Koblenz: ICPR, 1999.

———. *Ist der Rhein wieder ein Fluß für Lachs? "Lachs 2000."* Koblenz: ICPR, 1999.

———. *Internationale Vereinbarungen und Übereinkommen über den Schutz des Rheins gegen Verunreinigung.* Koblenz: ICPR, n.d.

International Energy Agency. *International Coal Trade: The Evolution of a Global Market.* Paris: OECD, 1997.

Intze, Otto. *Entwickelung des Thalsperrenbaues in Rheinland und Westfalen von 1889 bis 1903.* Düsseldorf, 1903.

Isenberg, Wolfgang, ed. *Historische Umweltforschung: Wissenschaftliche Neuorien-*

tierung—Aktuelle Fragestellungen. Bensberger Protokolle. Bergisch Gladbach: Thomas-Morus-Akademie Verlag, 1992.

Jahresbericht des Rheinischen Fischerei-Vereins für 1908/09. Bonn: Friedrich Cohen, 1909.

Jansen, P. Ph., L. van Bendegom, J. van den Berg, M. de Vries, and A. Zanen, *Principles of River Engineering: The Non-Tidal Alluvial River.* London: Pitman, 1979.

Jens, Günter. "Fische und Fischerei des Rheins." *Beiträge zur Rheinkunde,* no. 23 (1971): 8–14.

Jens, Günter, and Ragnar Kinzelbach. "Der Aal." *Mainzer Naturwissenschaftliches Archiv,* Beihefte 13 (1990): 69–74.

———. "Der Zander." *Mainzer Naturwissenschaftliches Archiv,* Beihefte 13 (1990): 75–77.

———. "Der Lachs." *Mainzer Naturwissenschaftliches Archiv,* Beihefte 13 (1990): 57–63.

Johann Gottfried Tulla. Karlsruhe: Theodor-Rehbock-Flußlaboratoriums (Karlsruhe University), 1970.

Jongman, Rob H. G. "Vegetation, River Management and Land Use in the Dutch Rhine Floodplains." *Regulated Rivers: Research and Management* 7 (1992): 279–89.

Juillard, Étienne. *L'Europe Rhénane: Géographie d'un grand espace.* Paris: Librairie Armand Colin, 1968.

Jurisch, Konrad Wilhelm. *Die Verunreinigung der Gewässer: Eine Denkschrift im Auftrage der Flusscommission des Vereins zur Wahrung der Interessen der chemischen Industrie Deutschlands.* Berlin: R. Gaertner's Verlagsbuchhandlung, 1890.

Kaiserliches Statistisches Amt. *Statistisches Jahrbuch für das Deutsche Reich.* Berlin: Puttkammer & Mühlbrecht, 1880.

Kellenbenz, Hermann. *Die Zuckerwirtschaft im Kölner Raum von der Napoleonischen Zeit bis zur Reichsgründung.* Cologne: Industrie- und Handelskammer zu Köln, 1966.

Keller, Reiner. *Natur und Wirtschaft im Wasserhaushalt der rheinischen Landschaften und Flußgebiete.* Forschungen zur Deutschen Landeskunde, vol. 57. Remagen: Verlag des Amtes für Landeskunde, 1951.

Kerner, Imre. *Der Rhein—Die Vergiftung geht weiter.* Reinbek bei Hamburg: Rowohlt, 1987.

Kinzelbach, Ragnar, ed. *Die Tierwelt des Rheins einst und jetzt: Symposium zum Jubiläum der Rheinischen Naturforschenden Gesellschaft und des Naturhistorischen Museums Mainz am 9. November 1984.* Mainz: Naturhistorisches Museum Mainz, 1985.

———. "Einschleppung und Einwanderung von Wirbellosen in Ober- und Mittelrhein (Coelenterata, Plathelminthes, Annelida, Crustacea, Mollusca)." *Mainzer Naturwissenschaftliches Archiv* 11 (1972): 109–50.

————. "Der Stör." *Mainzer Naturwissenschaftliches Archiv*, Beihefte 13 (1990): 51–56.

Kinzelbach, Ragnar, and G. Friedrich, eds. *Biologie des Rheins*. Limnologie Aktuell, vol. 1. Stuttgart: Fischer, 1990.

Kisch, Egon Edwin. *Mein Leben für die Zeitung 1926–1947: Journalistische Texte 2*. Vol. 9 of *Gesammelte Werke in Einzelausgaben*. Edited by Bodo Uhse and Gisela Kisch, 1983.

Klausewitz, Wolfgang. "Die frühere und heutige Fischfauna des Mains." In *Festschrift und 9. Bericht anläßlich des 100jährigen Bestehens des Fischereiverbandes Unterfranken e. V. 1877–1977*, 131–46. Würzburg: Fischereiverband Unterfranken, 1977.

Klein, Johann August. *Rheinreise von Mainz bis Köln: Historisch, topographisch, malerisch bearbeitet*. Koblenz: Fr. Röhling, 1828.

Kleinebeckel, Arno. *Unternehmen Braunkohle: Geschichte eines Rohstoffs, eines Reviers, einer Industrie im Rheinland*. Cologne: Rheinische Braunkohlenwerke Aktiengesellschaft, 1986.

Kleist, Heinrich von. *Erzählungen, Anekdoten, Gedichte, Schriften*. Vol. 3 of *Sämtliche Werke und Briefe*. Edited by Ilse-Marie Barth, Klaus Müller-Salget, Stefan Ormanns, and Hinrich C. Seeba. Frankfurt: Deutscher Klassiker, 1990.

Klomp, R. *A Model Approach to the Problem of Water Quality of the River Rhine*. Delft: Delft Hydraulics Laboratory, No. 256, 1981.

Kluge, Thomas, and Engelbert Schramm. *Wassernöte: Zur Geschichte des Trinkwassers*. Cologne: Volksblatt, 1988.

Knäble, K. "Tätigkeit und Werk Tullas." *Badische Heimat: Mein Vaterland* 50, no. 4 (1970): 450–65.

Kneese, Allen V., and Blair T. Bower. "Die Wasserwirtschaft im Ruhrgebiet: Eine Fallstudie der Genossenschaften." In *Umwelt und wirtschaftliche Entwicklung*, ed. Horst Siebert. Darmstadt: Wissenshaftliche Buchgesellschaft, 1979.

Knöpp, Herbert. "Der Rhein ein knappes Jahr nach 'Sandoz'." *Beiträge zur Rheinkunde*, no. 40 (1988): 5–19.

Koch, Karl-Wilhelm, and Gustav F. Röhr. *Der Rhein: Verkehrsweg im Herzen Europas Schienenwege und Schiffahrt*. Krefeld: Röhr-Verlag für Spezielle Verkehrsliteratur, 1985.

Koenig, Fritz. *Die Verhandlungen über die internationale Rheinregulierung im st. gallisch-voralbergischen Rheintal von den Anfängen bis zum schweizerisch-österreichischen Staatsvertrag von 1892*. Bern: Herbert Lang & Cie, 1971.

König, Josef. *Die Verunreinigung der Gewässer, deren schädliche Folgen nebst Mitteln zur Reinigung des Schmutzwassers*. Berlin, 1887.

Koepchen, Arthur. *RWE Elektrizitätswirtschaft*. Essen: RWE, 1930.

Köppen, Ursula von. "Die Geschichte der Kommission und die Rechtsordnung der Rheinschiffahrt." In *150 Jahre Zentralkommission für die Rheinschiffahrt*, 21–28. Duisburg-Ruhrort: Binnenschiffahrts-Verlag, 1966.

Kolb, Siegfried. "Naturschutz und Landschaftspflege bei wasserbaulichen Maßnahmen, dargestellt am Beispiel des Oberrheins." *Beiträge zur Rheinkunde*, no. 38 (1986): 5–18.

Kraats, J. A. van de, ed. *Rehabilitation of the River Rhine: Proceedings of the International Conference on Rehabilitation of the River, 15–19 March 1993, Arnhem, The Netherlands.* Oxford: Pergamon, 1994.

Kremer, Bruno. "Rheinaue und Auenvegetation im nördlichen Mittlerheingebiet." *Beiträge zur Rheinkunde*, no. 36 (1984): 38–49.

———. "Die Auengebiete des südlichen Oberrheins—Chance für die Natur oder verschwindende Naturlandschaft?" *Beiträge zur Rheinkunde*, no. 41 (1989): 56–67.

Kufferath-Sieberin, Günter. *Die Zuckerindustrie der linksrheinischen Bördenlandschaft.* Bonn: Selbstverlag des Geographischen Instituts der Universität Bonn, 1955.

Kunz, Egon. "Von der Tulla'schen Rheinkorrektion bis zum Oberrheinausbau." *Jahrbuch für Naturschutz und Landschaftsplege* 24 (1975): 59–78.

Kuster, Thomas. "Die Geschichte der internationalen Rheinregulierung und ihre Auswirkung auf die Bevölkerung und Wirtschaft des Vorarlberger Rheintals." Senior thesis, Innsbruck University, 1991.

La Navigation fluviale: Edition Spéciale de la Revue Schweizerland, July 1918.

Lambert, Audrey M. *The Making of the Dutch Landscape: An Historical Geography of the Netherlands.* London: Seminar Press, 1971.

Lamer, Mirko. *The World Fertilizer Economy.* Stanford: Stanford University Press, 1957.

Landes, David S. *The Unbound Prometheus: Technological Change and Industrial Development in Western Europe from 1750 to the Present.* Cambridge: Cambridge University Press, 1969.

Landesamt für Wasserwirtschaft Rheinland-Pfalz. *Rheinland-Pfalz. Gewässergüte Karten mit Erläuterungen.* Mainz: Ministerium für Umwelt und Gesundheit, 1988.

Landesanstalt für Umweltschutz Baden-Württemberg. *100 Jahre Gewässerkundliche Dienststelle Baden-Württemberg 1883–1983: Festschrift.* Karlsruhe: Landesanstalt für Umweltschutz Baden-Württemberg, 1984.

Lange, Jörg. "Robert Lauterborn (1869–1952)—Ein Leben am Rhein." *Lauterbornia*, no. 5 (September 1990): 1–18.

Langen, Rheinstrombaudirektor a. D. "100 Jahre Rheinstrombauverwaltung Koblenz." *Schriftenreihe des Rhein-Museum e. V. Koblenz* (1950).

Laspeyres, Renate. *Rotterdam und das Ruhrgebiet.* Marburger Geographische Schriften, no. 41. Marburg/Lahn: Marburg University, 1969.

Lauterborn, Robert. "Die Ergebnisse einer biologischen Probenuntersuchung des Rheins." *Arbeiten an der kaiserlichen Gesundheitsamt* 22 (1905): 630–52.

———. "Die Vegetation des Oberrheins." *Verh. naturhistorisch-medizinischen Vereins Heidelberg* N.F. 10 (1910): 450–502.

———. "Die geographische und biologische Gliederung des Rheinstromes." *Sitz-*

ungsbericht d. Heidelberger Akad. Wissensch. Math.-Nat. 5–7 (1916–18): 1–61, 1–70, and 1–87.

————. *Die erd- und naturkundliche Erforschung des Rheins und der Rheinlande vom Altertum bis zur Gegenwart.* Vol. 1 of *Der Rhein: Naturgeschichte eines deutschen Stromes.* Sonderabdruck aus den Berichten der Naturforschenden Gesellschaft zu Freiburg i. Br., 1–46, 1934.

Lelek, Anton, and Günter Buhse. *Fische des Rheins—früher und heute.* Berlin: Springer, 1992.

Les Actes du Rhin et de la Moselle. Strasbourg: Editions de la Navigation du Rhin, 1966.

Linse, Ulrich, Reinhard Falter, Dieter Rucht, and Winfried Kretschmer, eds. *Von der Bittschrift zur Platzbesatzung: Konflikte um technische Großprojekte.* Berlin: J. H. W. Dietz, 1988.

Livet, Georges, and Francis Rapp, eds. *Histoire de Strasbourg des origines à nos jours.* Strasbourg, 1980–82.

Löber, Ulrich, ed. *2000 Jahre Rheinschiffahrt.* Koblenz: Selbstverlag des Landesmuseums, 1991.

Löbert, Traude. *Die Oberrheinkorrektion in Baden: Zur Umweltgeschichte des 19. Jahrhunderts.* Mitteilungen des Institutes für Wasserbau und Kulturtechnik der Universität Karlsruhe. Karlsruhe: Institut für Wasserbau und Kulturtechnik, 1997.

Maffioli, Cesare S. *Out of Galileo: The Science of Waters, 1628–1718.* Rotterdam: Erasmus, 1994.

Malle, Karl-Geert. "Der Rhein: Modell für den Gewässerschutz." *Spektrum der Wissenschaft* (1983), 22–32.

Mantz, Gerhard. "Zur Erinnerung an Leben und Werk des Geheimen Regierungsrathes und Strombaudirektors Eduard Adolph Nobiling." *Beiträge zur Rheinkunde,* no. 34 (1982): 22–38.

McCully, Patrick. *Silenced Rivers: The Ecology and Politics of Large Dams.* London: Zed Books, 1996.

Meijers, A. "The Occurrence of Organic Micropollutants in the River Rhine and the River Maas in 1974." *Water Research* 10, no. 7 (1976): 597–604.

Melkonian, Michael, ed. *Ökologie des Rheins: Chancen und Risiken eines großen europäischen Stromes.* Bonn: Bouvier, 1992.

Mellor, Roy E. H. *The Rhine: A Study in the Geography of Water Transport.* O'Dell Memorial Monograph, no. 16. Aberdeen: University of Aberdeen, 1983.

Michels, Franz Xaver. "Die Entstehungsgeschichte des Rheins." *Beiträge zur Rheinkunde,* no. 25 (1973): 3–24.

Michels, Walter. *Unvergessene Dampfschiffahrt auf Rhein und Donau.* Darmstadt: Kommissionsverlag Hestra-Verlag, 1967.

Middelkoop, H., and C. O. G. van Haselen, eds. *Twice a River: Rhine and Meuse in the Netherlands.* Lelystad: National Institute for Inland Water Management and Waste Water Treatment (RIZA), 1999.

Ministerium für Umwelt Baden-Württemberg. *Hochwasserschutz und Ökologie: Ein 'Integriertes Rheinprogramm' schützt vor Hochwasser und erhält naturnahe Flußauen.* Stuttgart: Ministerium für Umwelt Baden-Württemberg, 1988.

Ministry of Transport, Public Works and Water Management. *Landscape Planning of the River Rhine in the Netherlands: Summary of the Main Report.* Delft: Directorate-General for Public Works and Water Management, 1996.

Miquel, Pierre. *Histoire des canaux, fleuves et rivières de France.* Paris: Edition 1, 1994.

Mordziol, C. "Zur Gliederung des Rheinstromes in einzelne Abschnitte." *Geographischer Anzeiger* 13 (1912): 231–32.

Morra, C. F. H. *Organic Chemicals Measured During 1978 in the River Rhine in the Netherlands.* Voorburg, 1979.

Morris, P. J. T., W. A. Campbell, and H. L. Roberts, eds. *Milestones in 150 Years of the Chemical Industry.* Cambridge: The Royal Society of Chemistry, 1991.

Moss, Brian. *Ecology of Fresh Waters: Man and Medium.* Oxford: Blackwell Scientific Publications, 1988.

Müller, Dieter, and V. Kirchesch. "On Nitrification in the River Rhine." *Verh. Internat. Verein. Limnol.* 22 (1985): 2754–60.

Mumford, Lewis. *Technics and Civilization.* New York: Harcourt, Brace and Company, 1934.

Musall, Heinz. *Die Entwicklung der Kulturlandschaft der Rheinniederung zwischen Karlsruhe und Speyer vom Ende des 16. bis zum Ende des 19. Jahrhunderts.* Heidelberger Geographische Arbeiten, no. 22. Heidelberg: Geographisches Institut der Universität Heidelberg, 1969.

Nelkin, Dorothy, and Michael Pollak. *The Atom Besieged: Antinuclear Movements in France and Germany.* Cambridge: MIT Press, 1982.

Nienhuis, P. H., R. S. E. W. Leuven, and A. M. J. Ragas, eds. *New Concepts for Sustainable Management of River Basins.* Leiden: Backhuys, 1998.

Niethammer, Günther. "Wandlungen in der Vogelwelt des Rheinlandes." *Rheinische Heimatpflege*, no. 3 (1964): 90–100.

Norlind, Arnold. *Die geographische Entwicklung des Rheindeltas bis um das Jahr 1500.* Lund: Hakan Ohlsson, 1912.

Nusteling, H. P. H. *De Rijnvaart in het tijdperk van stoom en steenkool (1831–1914).* Amsterdam: Holland Universiteits Pers, 1974.

Ott, Hugo, and Thomas Herzig. "Elektrizitätsversorgung von Baden, Württemberg und Hohenzollern 1913/14." In *Historischer Atlas von Baden-Württemberg*, vol. 11, Kommission für geschichtliche Landeskunde. Stuttgart: Landesvermessungsamt Baden-Württemberg, 1972–88.

Overbeck, Karl. "Die Wanderung der Großeisenindustrie des Ruhrgebiets zum Rhein." Ph.D. dissertation, Bonn University, 1923.

Pagee, J. *Water Quality Modelling of the River Rhine and Its Tributaries in Relation to Sanitation Strategies.* Delft: Delft Hydraulics Laboratory, no. 299, 1982.

Petts, G. E., H. Möller, and A. L. Roux, eds. *Historical Change of Large Alluvial Rivers: Western Europe.* Chichester: John Wiley, 1989.

Pichl, Karl. "Die Verbesserung des Schiffahrtsweges in der Binger-Loch-Strecke." *Beiträge zur Rheinkunde*, no. 12 (1961): 49–59.

Plumpe, Gottfried. *Die I. G. Farbenindustrie AG: Wirtschaft, Technik und Politik, 1904–1945*. Berlin: Duncker & Humblot, 1990.

Pohl, Hans, ed. *Gewerbe- und Industrielandschaften vom Spätmittelalter bis ins 20. Jahrhundert*. Vierteljahrschrift für Sozial- und Wirtschaftsgeschichte. Stuttgart: Steiner, 1986.

———. *Vom Stadtwerk zum Elektrizitätsgroßunternehmen: Gründung, Aufbau und Ausbau der "Rheinisch-Westfälischen Elektrizitätswerk AG" (RWE) 1898–1918*. Zeitschrift für Unternehmensgeschichte, no. 73. Stuttgart: Steiner, 1992.

Pohl, Hans, Ralf Schaumann, and Frauke Schönert-Röhlk. *Die chemische Industrie in den Rheinlanden während der industriellen Revolution: Die Farbenindustrie*. Zeitschrift für Unternehmensgeschichte. Wiesbaden: Steiner, 1983.

Porter, Cecelia Hopkins. *The Rhine as Musical Metaphor: Cultural Identity in German Romantic Music*. Boston: Northeastern University Press, 1996.

Ports du Rhin et Terrains Industriels de la Ville de Strasbourg. Strasbourg: Sociètè Alsacienne de Navigation Rhènane, 1923.

Pounds, N. J. G. *The Ruhr: A Study in Historical and Economic Geography*. New York: Greenwood, 1968.

———. *An Historical Geography of Europe*. Cambridge: Cambridge University Press, 1990.

Prager, Hans G. *Was weißt du vom Rhein?* Stuttgart: Franckh'sche Verlagshandlung, 1965.

Pretzel, Ulrike. *Die Literaturform Reiseführer im 19. und 20. Jahrhundert: Untersuchungen am Beispiel des Rheins*. Frankfurt: Peter Lang, 1995.

Przedwojski, B., R. Blazejewski, and K. W. Pilarczyk. *River Training Techniques: Fundamentals, Design and Applications*. Rotterdam: A. A. Balkema, 1995.

Pulp and Paper International. *The European Pulp and Paper Industry*. Brussels: Pulp and Paper International, 1991.

Quelle, Otto. *Industriegeographie der Rheinlande*. Institut für Geschichtliche Landeskunde der Rheinlande an der Universität Bonn, no. 5. Bonn: Kurt Schroeder, 1926.

Radcliffe, Ann. *A Journey, Made in the Summer of 1794 Through Holland and the Western Frontier of Germany with a Return Down the Rhine*. London: G. G. and J. Robinson, 1795.

Radkau, J. *Aufstieg und Krise der deutschen Atomwirtschaft, 1945–1975: Verdrängte Alternativen in der Kerntechnik und der Ursprung der nuklearen Kontroverse*. Reinbek bei Hamburg: Rowohlt, 1983.

Radkau, Joachim. *Technik in Deutschland: Vom 18. Jahrhundert bis zur Gegenwart*. Frankfurt: Suhrkamp, 1989.

Ramshorn, Alexander. "Die Wasserwirtschaft im rheinisch-westfälischen Industriegebiet." *Glückauf* (Sonderdruck) 88 (1952): 1–12.

———. *Die Emschergenossenschaft*. Essen: N. p., 1957.

Rees, Goronwy. *The Rhine.* New York: Putnam, 1967.

Reichelt, Günther. *Laßt den Rhein leben!* Düsseldorf: W. Girardet, 1986.

Rhein-Aktuell: Kurzinformation der internationalen Kommission zum Schutz des Rheins (IKSR).

Rheinurkunden: Sammlung zwischenstaatlicher Vereinbarungen, landesrechtlicher Ausführungsverordnungen und sonstiger wichtiger Urkunden über die Rheinschiffahrt seit 1803. Munich: Duncker & Humblot, 1918.

Ritter, Par Jean. *Le Rhin.* Paris: Presses Universitaires de France, 1963.

Römer, Gerhard. *Die Oberrheinlande in alten Landkarten: Vom Dreißigjährigen Krieg bis Tulla (1618–1828).* Karlsruhe: Selbstverlag der Badischen Landesbibliothek, 1981.

Rollins, William H. *A Greener Vision of Home: Cultural Politics and Environmental Reform in the German Heimatschutz Movement, 1904–1918.* Ann Arbor: University of Michigan Press, 1997.

Roth, Paul W. "Umweltschutzprobleme in den Jahren 1804 und 1890." *Blätter für Heimatkunde* 48 (1974): 26–30.

Royal Institute of International Affairs. *Regional Management of the Rhine: Papers of a Chatham House Study Group.* London: Chatham House, 1975.

Rüster, Bernd, and Bruno Simma, eds. *International Protection of the Environment: Treaties and Related Documents.* Dobbs Ferry, N.Y.: Oceana Publications, 1975.

Rutte, Erwin. *Rhein, Main, Donau: wie-wann-warum sie wurden.* Sigmaringen: Thorbecke, 1987.

Salomons, W., and A. J. de Groot. *Pollution History of Trace Metals in Sediments, as Affected by the Rhine River: Paper Presented at the Third International Symposium on Environmental Biogeochemistry, Wolfenbüttel, March 27–April 2, 1977.* Delft: Delft Hydraulics Laboratory, Publication No. 184, 1977.

Sander, Paul. "Die Internationale Kommission zum Schutze des Rheins gegen Verunreinigung." *Beiträge zur Rheinkunde,* no. 22 (1970): 14–18.

Sarkar, Saral. *Green-Alternative Politics in West Germany.* Tokyo: United Nations University Press, 1993–94.

Schaake, Hanns Dieter. *Der Fremdenverkehr in den linksrheinischen Kleinstädten zwischen Bingen und Koblenz.* In *Arbeiten zur Rheinischen Landeskunde.* Bonn: Ferd. Dümmlers, 1971.

Schachtner, Sabine. *Größer, schneller, mehr: Zur Geschichte der industriellen Papierproduktion und ihrer Entwicklung in Bergisch Gladbach.* Cologne: Rheinland-Verlag, 1996.

Schäfer Wilhelm. "Der Oberrhein als ökologisches Gefüge und seine ökotechnische Behandlung." *Jahrbuch für Naturschutz und Landschaftspflege* 24 (1975): 79–92.

Schaumann, Ralf. *Technik und technischer Fortschritt im Industrialisierungsprozeß: Dargestellt am Beispiel der Papier-, Zucker- und chemischen Industrie der nördlichen Rheinlande (1800–1875).* Rheinisches Archiv, vol. 101. Bonn: Ludwig Röhrscheid, 1977.

Schawacht, J. H. *Schiffahrt und Güterverkehr zwischen den Häfen des deutschen Niederrheins (insbesondere Köln) und Rotterdam vom Ende des 18. bis zur Mitte des 19. Jahrhunderts (1794–1850/51).* Schriften zur rheinisch-westfälischen Wirtschaftsgeschichte, vol. 26. Cologne: Rheinisch-westfälisches Wirtschaftsarchiv zu Köln, 1973.

Schenck, Ulrich. "Die Wasserwirtschaft im Niederschlagsgebiet der Ruhr: Eine volkswirtschaftliche Untersuchung." Ph.D. dissertation, Cologne University, 1931.

Schiller, Friedrich. *Werke. Nationalausgabe.* Vols. 1–30. Edited by Julius Petersen and Gerhard Fricke. Weimar: H. Böhlaus Nachfolger, 1943.

Schlegel, F. "Zur naturnahen Neugestaltung des Alpenrheins." *Werdenberger Jahrbuch* 3 (1990): 184–92.

Schmelcher, Ernst. *RWE, 1898–1954.* Essen: RWE, 1954.

Schmidt, Hans M., ed. *Der Rhein-le Rhin-de Waal: Ein europäischer Strom in Kunst und Kultur des 20. Jahrhunderts.* Cologne: Wienand, 1995.

Schmidt-Ries, Hans. *Limnologische Untersuchungen des Rheinstromes.* Vol. 4. Opladen: Westdeutscher Verlag, 1973.

Schmitz, Walter, ed. *50 Jahre Rhein-Verkehrs-Politik.* Duisburg: "Rhein" Verlagsgesellschaft m.b.H., 1927.

Schneider, Helmut J., ed. *Der Rhein: Seine poetische Geschichte in Texten und Bildern.* Frankfurt: Insel, 1983.

Schneider, Matthias. *Wasserhaushalt und Wasserwirtschaft im Gebiete der Erftquellflüsse (Nordeifel).* Bonn: Selbstverlag des Geographischen Instituts der Universität Bonn, 1953.

Schnitter, Niklaus. *Geschichte des Wasserbaus in der Schweiz.* Oberbözberg: Olynthus, 1992.

Schöttler, Peter. "The Rhine as an Object of Historical Controversy in the Inter-War Years. Towards a History of Frontier Mentalities." *History Workshop Journal,* no. 39 (Spring 1995): 1–21.

Schultz, Friedrich. *Deutscher Braunkohlen-Industrie-Verein e. V. 1885–1960.* Düsseldorf: Braunkohle, n.d.

Schultze-Rhonhof, Friedrich-Carl. *Die Verkehrsströme der Kohle im Raum der Bundesrepublik Deutschland zwischen 1913 und 1957: Eine wirtschaftsgeographische Untersuchung.* Geographisches Institut der Eberhard-Karls-Universität zu Tübingen, vol. 146. Bad Godesberg: Bundesanstalt für Landeskunde und Raumforschung, 1964.

Schwab, Prof. Dr. "Der Naturschutz im Rheinstromgebiet." *Beiträge zur Rheinkunde,* no. 4 (1928): 31–34.

Schwarzmann, Herbert. "War die Tulla'sche Oberrhein-korrektion eine Fehlleistung im Hinblick auf ihre Auswirkungen?" *Die Wasserwirtschaft* 54, no. 10 (1964): 279–87.

Schweng, Erich. "Der amerikanische Flußkrebs Orconectes Limosus (Rafinesque) im Rhein." *Mainzer Naturwissenschaftliches Archiv* 7 (1968): 265–74.

Scott, James C. *Seeing Like a State: How Certain Schemes to Improve the Human Condition Have Failed.* New Haven: Yale University Press, 1998.

Seifert, Alwin. "Die Versteppung Deutschlands." *Deutsche Technik* 4 (September 1936): 423–27.

Shelley, Mary. *Frankenstein; or, the Modern Prometheus.* 1818. Berkeley and Los Angeles: University of California Press, 1994.

Sieferle, Rolf Peter, ed. *Forschritte der Naturzerstörung.* Frankfurt: Suhrkamp, 1988.

———. *Der unterirdische Wald. Energiekrise und industrielle Revolution.* Munich: C. H. Beck, 1982.

Sigmond, Johannes. "Landschaftswandel und Wasserhaushalt im rheinischen Braunkohlengebiet." In *Das Rheinische Braunkohlengebiet: Eine Landschaft in Not!* Rheinischer Verein für Denkmalpflege und Heimatschutz. Neuss: Gesellschaft für Buchdruckerei, 1953.

Simson, John von. "Die Flußverunreinigungsfrage im 19. Jahrhundert." *Vierteljahrschrift für Sozial- und Wirtschaftsgeshichte* (1978), 370–89.

Singer, Charles, ed. *A History of Technology.* Oxford: Clarendon, 1954–84.

Smith, Norman. *Man and Water: A History of Hydro-Technology.* London: Charles Scribner's Sons, 1975.

Smith, Robert Angus. *Air and Rain: The Beginnings of Chemical Climatology.* London: Longmans, Green & Co., 1872.

Solmsdorf, Hartmut, Wilhelm Lohmeyer, and Walter Mrass. *Ermittlung und Untersuchung der schutzwürdigen und naturnahen Bereiche entlang des Rheins (Schutzwürdige Bereiche im Rheintal).* Bonn–Bad Godesberg: Bundesanstalt für Vegetationskunde, Naturschutz und Landschaftspflege, 1975.

Solmsdorf, Hartmut and Antje. "Schutzwürdigen Bereiche im Rheintal." *Beiträge zur Rheinkunde,* no. 27 (1975): 28–31.

Spelsberg, Gerd. *Rauchplage. Hundert Jahre Saurer Regen.* Aachen: Alano, 1984.

Stein, G. *Stadt am Strom: Speyer und der Rhein.* Speyer: Verlag der Zechner'schen Buchdruckerei in Speyer, 1989.

Stenographische Berichte über die Verhandlungen des Reichstages. Berlin: Verlag der Norddeutschen Buchdruckerei, 1871–1933.

Stenz, W. "Zu den Beziehungen zwischen Gesellschaft und Umwelt von der Industriellen Revolution bis zum Übergang zum Imperialismus." *Jahrbuch für Wirtschaftsgeschichte* 1 (1984): 81–132.

Stokes, Raymond G. *Opting for Oil: The Political Economy of Technological Change in the West German Chemical Industry, 1945–1961.* Cambridge: Cambridge University Press, 1994.

Stottrop, Ulrike, ed. *Zeitraum Braunkohle.* Essen: Ruhrlandmuseum, 1993.

Strack, H. R. "Der Hochrhein von Basel bis zum Bodensee." *Beiträge zur Rheinkunde,* no. 11 (1960): 27–37.

Strasser, Rudolf. *Die Veränderungen des Rheinstromes in historischer Zeit.* Düsseldorf: Droste, 1992.

Stroud, Richard H., ed. *Fish Culture in Fisheries Management*. Bethesda: American Fisheries Society, 1986.

Tacitus, Publius Cornelius. *The Annals*. The Loeb Classical Library. Cambridge: Harvard University Press, 1955.

Taylor, F. Sherwood. *A History of Industrial Chemistry*. New York: Abelard-Schuman, 1957.

TeBrake, William H. *Medieval Frontier: Culture and Ecology in Rijnland*. College Station: Texas A & M University Press, 1985.

Tennant, Charles. *A Tour through Parts of the Netherlands, Holland, Germany, Switzerland, Savoy, and France in the Years 1821–1822*. London: Longman, 1824.

Teuteberg, Hans-Jürgen, ed. *Westfalens Wirtschaft am Beginn des "Maschinenzeitalters."* Dortmund: Gesellschaft für Westfälische Wirtschaftsgeschichte, 1988.

Thienemann, August. "Hydrobiologische und fischereiliche Untersuchungen an den westfälischen Talsperren." *Landwirtschaftliche Jahrbücher* 41 (1911): 535–716.

———. "Die Verschmutzung der Ruhr im Sommer 1911." *Zeitschrift für Fischerei und deren Hilfswissenschaften* 16 (1912): 55–86.

Thoreau, Henry David. "Walking." In *The Works of Henry D. Thoreau: Excursions*. New York: Thomas Y. Crowell Company, 1913.

Tippner, Manfred. "Der Feststofftransport am Rhein." *Beiträge zur Rheinkunde*, no. 27 (1975): 32–38.

———. "Schwebstoffmessungen im Rhein vor 125 Jahren." *Beiträge zur Rheinkunde*, no. 31 (1979): 67–68.

Tittizer, Thomas, and Falk Krebs, eds. *Ökosystemforschung: Der Rhein und seine Auen: Eine Bilanz*. Berlin: Springer, 1996.

Tittizer, T., F. Schöll, M. Dommermuth, J. Bäthe, and M. Zimmer. "Zur Bestandsentwicklung des Zoobenthos des Rheins im Verlauf der letzten neun Jahrzehnte." *Wasser und Abwasser* 35 (1991): 125–66.

Tittizer, T., F. Schöll, and D. Hardt. "Die Lebensgemeinschaft der Rheinsohle." *Beiträge zur Rheinkunde*, no. 44 (1992).

Tittizer, Thomas, Franz Schöll, and Maria Dommermuth. "Die Entwicklung der Lebensgemeinschaften des Rheins im 20. Jahrhundert." In *Die Biozönose des Rheins im Wandel: Lachs 2000? Beiträge der Fachtagung am 26. März 1992 im Kurfürstlichen Schloß in Mainz*. N.p.: Ministerium für Umwelt Rheinland-Pfalz, 1992.

Tittizer, Thomas, and Franz Schöll. "Leben an der Stromsohle des Rheins. Untersuchungen mit Taucherschacht und Taucherglocke." *Biologie in unserer Zeit* 23, no. 4 (1993): 248–53.

Topic Rhine: Latest News from the International Commission for the Protection of the Rhine.

Trahms, Oberregierungsrat. "Der Rhein: Abwasserkanal oder Fischgewässer? Eine Betrachtung über die Fischereiverhältnisse im Rhein." *Beiträge zur Rheinkunde*, no. 7 (1955): 43–56.

Travis, Anthony S. *The Rainbow Makers: The Origins of the Synthetic Dyestuffs Industry in Western Europe.* Bethlehem: Lehigh University Press, 1993.

———. "Poisoned Groundwater and Contaminated Soil: The Tribulations and Trial of the First Major Manufacturer of Aniline Dyes in Basel." *Environmental History* 2 (July 1997): 343–65.

Treue, Wilhelm. *150 Jahre Köln-Düsseldorfer: Die Geschichte der Personenschiffahrt auf dem Rhein.* Cologne: Köln-Düsseldorfer Deutsche Rheinschiffahrt AG, 1976.

Tümmers, Horst-Johannes. *Rheinromantik: Romantik und Reisen am Rhein.* Cologne: Greven, 1968.

———. *Der Rhein: Ein europäischer Fluß und seine Geschichte.* Munich: C. H. Beck, 1994.

Tulla, J. G. *Über die Rektifikation des Rheins, von seinem Austritt aus der Schweiz bis zu seinem Eintritt in das Großherzogthum Hessen.* Karlsruhe: Chr. Fr. Müller's Hofbuchdruckerei, 1825.

Twain, Mark. *A Tramp Abroad.* In *The Writings of Mark Twain.* Definitive Edition, vol. 9. New York: Gabriel Wells, 1923.

Uhde, Regierungsbaudirektor. "Die derzeitige Verschmutzung von Bodensee, Hoch- und Oberrhein." *Beiträge zur Rheinkunde*, no. 7 (1955): 29–42.

Veen, Johan van. *Dredge, Drain, Reclaim: The Art of a Nation.* The Hague: Martinus Nijhoff, 1962.

Ven, G. P. van de, ed. *Man-Made Lowlands: History of Water Management and Land Reclamation in the Netherlands.* Utrecht: Matrijs, 1993.

———. *Aan de wieg van Rijkswaterstaat: Woordingsgeschiedenis van het Pannerdenskanaal.* Zutphen: De Walburg Pers, 1976.

Versell, W., and Ant. Schmid. *Bericht über Wildbachverbauungen im bündnerischen Rheingebiet zur Sicherung der Rheinregulierung oberhalb des Bodensees.* Chur, Switzerland: Buchdruckerei A.-G. Bündner Tagblatt, 1928.

Vetter, G. "Die Pflanzengesellschaften am Roxheimer Altrhein." *Pfälzer Heimat, Speyer* 27, no. 2 (1976): 49–56.

Vondel, Joost van den. *De Werken van Vondel.* Amsterdam: Maatschappij voor Goede en Goedkoope Lectuur, 1929.

Wanner, Gerhard. *Vorarlbergs Industriegeschichte.* Feldkirch: Verein Vorarlberger Industriegeschichte, 1990.

Wasser- und Schiffahrtsdirektion Duisburg. *Der Rhein: Ausbau, Verkehr, Verwaltung.* Duisburg: "Rhein" Verlagsgesellschaft m.b.H., 1951.

"Wasserwirtschaftlicher Verband. Hauptversammlung, Berlin, 24. 2. 1912." *Zeitschrift für angewandte Chemie* 25 (26 April 1912): 835–38.

Wasserwirtschaftsverwaltung des Landes Nordrhein-Westfalen und der Wasser- und Schiffahrtsdirektion Duisburg. *Die Verunreinigung des Rheins im Lande Nordrhein-Westfalen (Stand: Ende 1956): Maßnahmen zu ihrer Bekämpfung.* Düsseldorf: Wasserwirtschaftsverwaltung des Landes Nordrhein-Westfalen und der Wasser- und Schiffahrtsdirektion Duisburg, 1958.

Weber, Heinz. *Die Anfänge der Motorschiffahrt im Rheingebiet*. Duisburg-Ruhrort: Binnenschiffahrts-Verlag, 1978.

Weigelt, Curt. "Die Industrie und die preussische Ministerialverfügung von 20. Februar 1901." *Die Chemische Industrie* 24 (15 October 1901): 555–59.

Weimann, Reinhold. *Regulierungen und Kanalisierungen unserer Flüsse, biologisch gesehen*. Heisterbacherott: Selbstverlag, 1964.

Welcomme, Robin L. *Fisheries Ecology of Floodplain Rivers*. London: Longman, 1979.

Wey, Klaus-Georg. *Umweltpolitik in Deutschland: Kurze Geschichte des Umweltschutzes in Deutschland seit 1900*. Opladen: Westdeutscher Verlag, 1982.

White, Richard. *The Organic Machine: The Remaking of the Columbia River*. New York: Hill and Wang, 1995.

Whitton, B. A., ed. *Ecology of European Rivers*. Oxford: Blackwell Scientific Publications, 1984.

Wiebeking, Karl Friedrich von. *Atlas enthaltend drey und dreyssig hydrographisch-tophographische Karten von dem groesten Theil des schiffbaren Rheins*. Munich, 1832.

Wiebenga, Ben. "Pieter Calands Plan." *Beiträge zur Rheinkunde*, no. 24 (1972): 19–24.

Wille, Ulrich. "Die Brutvögel zweier Baggerlöcher am unteren Niederrhein." *Rheinische Heimatpflege*, 2 (1966): 139–53.

Williams, Raymond, ed. *The Pelican Book of English Prose*. Baltimore: Pelican, 1969.

Wittmann, H. "Tulla, Honsell, Rehbock." *Bautechnik-Archiv*, no. 4 (1949): 4–52.

Zängl, Wolfgang. *Deutschlands Strom: Die Politik der Elektrifizierung von 1866 bis heute*. Frankfurt: Campus, 1989.

Zier, H. G. "Johann Gottfried Tulla. Ein Lebensbild." *Badische Heimat* 50, no. 4 (1970): 379–449.

Zimmermann, Josef. *Landschaft verhandelt—mißhandelt: Mensch und Umwelt in nordrheinischen Ballungszentren*. Edited by Rheinischer Verein für Denkmalpflege und Landschaftsschutz e. V. Neuss: Neusser Druckerei und Verlag, 1982.

Index

Aachen, 100

Aare river: geophysical features of, 26; re-engineering of, 64; nuclear plants on, 135; mentioned, 12, 21, 69, 72, 142

Aargau canton, 26

Acid industry, 112–17, 119. *See also* Sulfuric acid

Acid rain, 77, 120

Agfa, 116–17

Agriculture: threats to Baden's water table, 66–67; sugar-beet production, 125–26; fertilizer runoff, 148–49, 184; political influence of Alsatian farmers, 179–80; putting an end to reclamation projects, 195–96; mentioned, 153–54, 156, 206. *See also* Fertilizers

Ahr river, 156

Alkali industry, 112–17, 120

Allis shad. *See* Shad

Alpenrhein Project, 62–63

Alpenrhein river: geophysical features of, 24–26; re-engineering of, 62–63, 69; hydroelectric production on, 131–33, 142; fish species found in, 160; mentioned, 12. *See also* Hinterrhein river; Vorderrhein river

Alps. *See* Aare river; Alpenrhein river; Switzerland

Alsace. *See* Alsace-Lorraine

Alsace-Lorraine: and Upper Rhine, 27–28; and Tulla Project, 50–55

passim; and Grand Canal d'Alsace, 66–68; and steel production, 86–87; and potash mining, 114, 149; and hydroelectric development, 133–34; and nuclear energy 135–37; and chloride pollution, 179–80; and salmon restoration, 189–90; mentioned, 4, 14, 23, 40, 41, 194. *See also* Upper Rhine

Aluminum industry, 132

Amazon basin, 21

Amphibians, 155–56

Amsterdam-Rhine canal, 30, 62

André, Fritz, 18, 69–70, 72, 74

Arndt, Ernst Moritz, 9–10

Arsenic poisonings, 121–23

Association of Rhine and Meuse Water Supply Companies, 178

Association Saumon-Rhin, 190

Austria: and re-engineering of Alpenrhein, 62–63; hydroelectric development, 131–33; mentioned, 4, 33, 178

Aventis, 23, 142. *See also* Hoechst

Bacharach, 7, 17, 28–29

Baden: and Tulla Project, 49–54, 65–68; and flood problems from re-engineering, 70–72, 74; and pollutants from chemical industry, 140; mentioned, 23, 40, 41, 128, 133, 173. *See also* Baden-Württemberg